Understanding te

LIST OF TABLES

UNDERSTANDING TENDERING AND ESTIMATING

A. A. KWAKYE MSc, ARICS, AGIS, MCIOB

Gower

Published by
Gower Publishing Limited
Gower House
Croft Road
Aldershot
Hampshire GU11 3HR
England

Gower
Old Post Road
Brookfield
Vermont 05036
USA

A. A. Kwakye has asserted his right under the Copyright, Designs and Patents Act 1988 to be identified as the author of this work.

British Library Cataloguing in Publication Data

Kwakye, A. A.
  Understanding Tendering and Estimating
  I. Title
  692.5

  ISBN 0-566-07490-7 Hardback
       0-566-07509-1 Paperback

Library of Congress Cataloging-in-Publication Data

Kwakye, A. A.
    Understanding tendering and estimating / A.A. Kwakye.
       p. cm.
    Includes bibliographical references.
    ISBN 0-588-07490-7 – ISBN 0-588-07509-1 (pbk.)
    1. Building–Estimates–Great Britain.   2. Contracts for work and
  labor–Great Britain.   I. Title.
  TH435.K93   1994
  692'.5'0941–dc20   93-48704
                                                                     CIP

Typeset in New Century Schoolbook by Manton Typesetters, 5–7 Eastfield Road, Louth, Lincolnshire and printed in Great Britain by Hartnolls Ltd, Bodmin.

CONTENTS

v

LIST OF FIGURES

xi

FOREWORD

Having considered this book and its title I turned for reference to a dictionary of building and found the definition of estimating as: 'Determining the probable cost of work by multiplying the volume of different operations by cost per unit of measurement known from recent work', and for tendering: 'An offer from a contractor to do certain work for a price which he or she names, usually in a priced bill'. Simple enough explanations but this book correctly explores this simplicity in detail.

Tendering and estimating are skills that are as old as trade itself but, as time goes by, the external influences upon them become apparently more complex and diverse. This is especially true of the construction industry and particularly of the influence of changes to the process which have occurred rapidly over the last ten years. The systems of procurement available in the marketplace are diverse, and each with their own peculiarities posing many questions and considerations for the estimator and tenderer as well as for the client or user and his or her professionals.

This book sets out in clear and understandable terms the alternative systems and approaches which can be applied throughout the whole process. The section on estimating demonstrates how the build-up of unit rates is undertaken, which can form the basis for an estimator's reference bible that will be built up over the years.

The reader, whether experienced or a student, will find this book most useful in developing a fuller understanding of the construction process and, in particular, the tendering and estimating skills relating to it.

L. J. Yadoo, FCIOB MAPM

PREFACE

Construction is a complicated process that embraces many activities and participants both on and off site. The process also involves expenditure of money, price determination and, in most cases, competition. In a competitive world, tendering and estimating are of importance to clients as they facilitate selection of contractors, and to contractors as the only means of securing construction business.

This volume has been written with the aim of providing students on degree courses in Construction, Construction Management, Quantity Surveying, Building Surveying, Architecture, and also students on TEC Level 4, Estimating, Quantity Surveying, Building Surveying and Construction Management courses, with an appreciation of tendering and estimating for construction work.

The main themes developed are fourfold: first, clients' development strategy, which influences the selection of a suitable procurement method and contractor for the achievement of their development goals; secondly, the efficient utilization of contractor's resources and the consequential submission of competitive bids; thirdly, the complexities of tendering and estimating; and finally, tender analysis and direction.

After a general section on the construction industry in Chapter 1, Chapters 2 and 3 are devoted to the examination of the client's development aims and modes of contractor selection. Chapter 4 concerns analysis of the contractor's business objectives. Chapters 5, 6 and 7 set out to explore tendering and estimating processes with the aid of many worked examples, while Chapter 8 deals with tender analysis, correction of errors, current trends and tendering arrangements outside the UK.

In addition to the estimating information contained within this text, students should endeavour to familiarize themselves with the National Working Rule Agreement for the Building Industry published by the National Joint Council for the Building Industry.

A. A. Kwakye

ACKNOWLEDGEMENTS

The author expresses his thanks to Ben Beke BSc, MSc, ARICS, AGIS, MCIOB; Raymond Floyd MCIOB; Paul Kabenla; George Mends BSc; Richmond Oppong MSc, MCIOB; and Howard Randall for the most useful advice and assistance in various forms, comments and encouragement he has received during the preparation of this book.

Grateful thanks are also due to Robert Fraser BSc for his outstanding work in correcting the original text and for invaluable suggestions that have enhanced the value and quality of the book. In addition, the author has benefited immeasurably from Adrian A. B Wood ARICS for reading the draft, for his constructive criticism and invaluable suggestions/comments, and for correcting some sections of the book.

The author is grateful also to Florette McCammon for her patience and splendid work of typing the manuscript, and to his wife Tina for her continued assistance and support.

CONSTRUCTION PROCUREMENT

CONSTRUCTION GENERALLY

Construction activity worldwide is a means of altering the environment by the production, removal or refurbishment of a unit of construction project. Conceptually, this activity takes place under the effect of cause, space and time and involves the active participation of several individuals each with separate responsibilities. The initiation of a construction activity invokes the expenditure of money, the employment of labour and machine power, and the management of time and resources. It is therefore important to ensure that these resources are readily available before embarking on any construction project.

TYPES OF CONSTRUCTION ACTIVITIES

Construction activities may be classified as new work or rehabilitation/maintenance.

New work
New construction works comprise construction products built on virgin sites as an addition to the existing stock. Similarly, new works may be constructed in a built environment as extensions to the existing stock or to replace those redundant.

Rehabilitation/maintenance

If the existing stock of buildings become unsuitable for their present use, then improvement, repair and maintenance are often required to keep them in usable condition. Work under this classification may involve substantial change of use. An improvement of the existing stock may be undertaken to retain the value of the investment, present a good appearance and raise standards to levels acceptable to the community.

 ## DEMAND FOR CONSTRUCTION PRODUCTS

The demand for construction products can be direct as well as derived. *Direct demand* occurs when the product is required for direct consumption. The types of construction product which fall into this category include houses, swimming pools, cinemas, car parks, and educational, church and hospital buildings. The demand for construction products, when described as *derived demand*, expresses product for production process, that is, product wanted because of its contribution to the production of other consumer goods and services. Construction products that correspond to this description include factories, warehouses, roads, railways and office buildings.

The demand for construction products varies and depends greatly on the economic activities in a country. The demand is met when a person or an organization recognizes user potential for the product and therefore initiates and completes the production process. Thus building needs to be initiated, that is, set in motion, and the initiator is known as the client or the employer.

 ## THE CONSTRUCTION INDUSTRY

In the United Kingdom, the production of construction products takes place within an all-important industry, the construction industry.

The construction industry is the industry responsible for the procurement of investment products for promoting production process or for direct consumption. It is an industry which can be best described as a collection of industries, because a building is composed of an assembly of building materials/components and equipment produced

Table 1.1   Stages of construction activity

Conception	The building owner, having decided to develop a piece of land, appoints an architect of his or her choice. He or she briefs the architect of his or her user requirement and cost limit (i.e. the amount of capital available for the project).
Design	The appointed architect explores the feasibility of the building owner's proposal. He or she then carries out the process of designing a building that meets the building owner's requirements in terms of accommodation, cost, quality and time.
Documentation	The building owner's professional advisers select the type of contract that suits the project in terms of cost, time and general market conditions. Contract documentation is prepared to aid selection of a suitable builder and the execution of the construction works.
Tendering and estimating	Completed documents are sent to selected builders to enable them to submit competitive bids for the project. In most cases, the builder who submits the lowest tender price is awarded the contract.
Construction	The selected builder undertakes the construction of the building to the shape, size and quality as depicted on the architect's drawings and specification.
Commissioning	The completed building is handed over to the building owner and the architect ensures that the building and its services perform to the owner's expectations. The architect gives the owner general guidance on maintenance and as-built drawings for buildings, drainage and service installations.

by other industries. Examples of products from the construction industry include buildings (houses, offices, factories, warehouses, hospitals) which provide shelter and contribute to the well-being and functioning of a modern national economy.

The primary tasks in the UK construction process may be arranged in the following order:

- Conception
- Design
- Documentation
- Production
- Commissioning

These distinct, but interrelated, processes associated with the projects are constrained by time, resources and performance. They also involve a wide range of individuals with practical and professional skills.

As activities in Table 1.1 show, various participants perform different functions in each phase of the construction process. Every effort is made by the participants to solve all problems posed by the project and endeavour to achieve the building owner's aims. (See Table 1.2 for participants of the construction process.)

CHARACTERISTICS OF THE CONSTRUCTION INDUSTRY

There are certain peculiar characteristics which are attributed to the UK construction industry. The reasons for the peculiar mode of operations in the construction industry can be best explained as the result of its size, diversity of operations in wide geographical areas, organization of production process, price determinant procedures, number of specialist firms required and various sources of equipment, material and components for production. The peculiar characteristics may be listed as follows:

1. The industry is fragmented in terms of the number of institutions representing construction professionals, the number of trade associations representing tradesmen, and specialization within the industry.
2. Design is separated from construction (i.e. production begins when design process ends).

Table 1.2 Construction process and participants

Participants	Construction process					
	Conception	Design	Contract documentation	Tendering and estimating	Construction	Commission
Client	▲				○	○
Architect		▲	○	○	○	▲
Quantity surveyor		○	▲	○	○	○
Structural engineer		▲	○		○	○
Services engineer		▲	○		○	○
Contractor		■	■	▲	▲	○
Domestic subcontractor					▲	
Specialist subcontractor					▲	
Statutory authority		○			○	○
Project manager	■	■	■	■	■	■

Note: ▲ key participants; ■ participated under integrated system; ○ passive participant

3. Employment of casual labour and subcontracting instead of full-time employed operatives.
4. Lack of investment in fixed capital assets and reliance on sub-contracting and plant hire.
5. The amount of work available to the industry and its firms over a period of time is difficult to predict. This makes it impossible to forecast the training needs, the level of investment and the level and value of each firm's output.
6. The industry has a complex structure in terms of the different types of contractors, specialist subcontractors and professional firms.

 ## CLASSIFICATION OF CLIENTS

The construction industry looks for work in many spheres of activity. However, the industry's clients may be classified generally as public sector clients or clients from the private sector.

Public sector clients

These are public authorities whose operations are governed generally by Acts of Parliament. They operate as agents of the central government which exercises control over their capital building expenditure. Thus the amount of construction activity these bodies can undertake depends on direct central government spending on capital investment. A cut in central government spending may mean postponement of expenditures on major schemes such as motorways, schools, housing and so on. The public authorities need the approval of the central government to undertake certain projects even if they can raise the necessary finances for the project. The public sector clients include the following:

1. Central government departments: for example, the Department of the Environment and the Property Services Agency who are responsible for all central government works.
2. Local authorities who are responsible for the provision of housing, schools, libraries, swimming pools, halls, sports centres and so forth.
3. Health authorities who are responsible for hospital buildings.
4. Public corporations, (for example, the National Coal Board, British Railways and Air Transport boards) each responsible for the

provision of buildings and other construction products required for its own particular field of operation.

Private sector clients

These are private companies which build for leasing, sale or own occupation. The central government exercises a limited amount of control over their operations which takes the form of acquisition of planning permission for their building programmes.

Private sector clients may be classified as follows:

1. Multinational companies: for example, Ford, Shell, ICI and Esso, whose buildings include factories, production plants, offices and distribution depots.
2. National companies: for example, Tescos, Sainsburys and Marks & Spencer, who construct buildings for their own use in warehousing and retail.
3. Local companies who construct offices, factories, shops and houses for their own use or sale.
4. Private clients who construct new buildings and also extend, refurbish or repair existing buildings for their own occupation, letting or sale.

 PARTICIPANTS IN THE CONSTRUCTION PROCESS

The successful production of any construction product depends on the combined efforts of several participants, as follows:

1. The client who is an individual or an organization usually initiates the construction process by commissioning various construction professionals to build to specific requirements.
2. Range of construction professionals include the project manager, architect, quantity surveyor, structural engineer and the services engineer.
3. The construction team is composed of the main contractor, specialist firms and material suppliers.
4. The material manufacturer/suppliers and the plant hire firms.
5. Public authorities who ensure that the various building regulations are complied with and that public health and safety regulations are observed. These are officers from the local authority, safety executive, fire, water, gas and electricity boards.

6.   The legal profession which oversees construction contractual matters.
7.   The financiers, that is, persons lending money (for example, banks, financial institutions and insurance companies).
8.   The ultimate consumer, who is either the owner or tenant.

MANAGEMENT METHODS FOR THE CONSTRUCTION PROCESS

Management methods may be classified in respect of the level of integration of the design and construction process as either traditional system or the integrated system.

Traditional system

In terms of timing and responsibility under this approach, design is separated from construction and each stage of the production process managed separately. This management method may be characterized as a sequential approach: conception, development and implementation phases are each completed and approved before proceeding to the next (see Table 1.1).

The separation of design and production functions (shown in Figure 1.1), is prone to lead to design without concern for buildability or production economies, a lack of feedback to the design team, and perpetuation of mistakes from project to project.

Integrated system

This is a generic term for the many systems which seek to integrate the design and construction processes, while retaining the separation of responsibilities. This system allows the contractor to play an active part in design as well as in construction.

PRICE DETERMINATION

Generally the price clients pay for their development projects should be fair and reasonable in relation to the amount of work done. Therefore prices in the construction industry are usually determined by

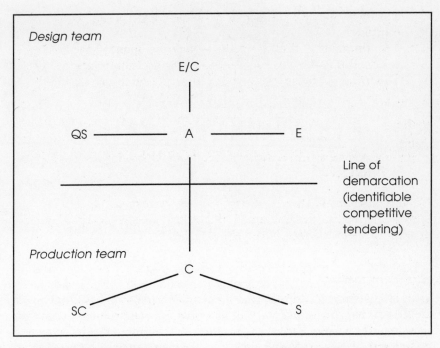

Key:
E/C	Employer/client	E	Engineer (structures/services)
A	Architect	C	Contractor
QS	Quantity surveyor	SC	Subcontractor
S	Supplier		

Figure 1.1 Typical separation of design and production functions

competitive tendering to effect a high level of activity. However, some private sector clients find it advantageous to employ negotiated tendering mechanisms in their price determination.

THE CLIENT'S DEVELOPMENT STRATEGY

 THE CLIENT'S ECONOMIC CONSIDERATIONS

All development projects arise from user or potential user demand for the building. The client, recognizing this demand, undertakes a development for sale, letting, leasing or own occupation. Whatever his or her motive, the client will make every effort to achieve value for money for his or her building programme. Therefore the professional advisers will adopt the following plan of action to acquire a product of high quality, completed on time at an optimum cost:

1. Selection of an appropriate procurement method in line with the project needs.
2. Clear and adequate design brief embracing all the requirements of the project.
3. High-standard design which promotes buildability, and ensures that the contractor's resources are used to the best advantage.
4. In the case of projects for client's own occupation, selection of materials and components to reflect the effects of maintenance or repair expenditure during the life of the building.
5. Exercise of effective cost control during design and construction phases of the development in order to maintain building cost within client's development budget.
6. Supply of adequate construction information when needed to avoid disruption of the contractor's progress and eventually, overrun on the construction programme.
7. Avoidance of variations to the works and where unavoidable, kept to a minimum.
8. Development planned ideally to take advantage of the weather (e.g. groundworks planned to take place in summer in lieu of winter saves time and money).

TRAITS OF A SUITABLE CONTRACTOR

A good contractor for the project is an important factor contributing to its completion and the client's professional advisers may look for the following attributes when drawing up a tender list:

Contractor's reputation in business

The contractor's record of performance in the construction business: that is, the number of projects he or she has completed successfully and sound financial standing are the attributes of a reputable firm.

Contractor's potential resources

This embraces physical as well as human resources. For example: buildings, offices, workshops, factories, fixed plant and machinery; the number and types of trade operatives regularly employed and the quality of management personnel, and their technical knowledge and experience both in the office and on site.

Contractor's normal conduct of business

Considerations under this heading include: the type of work package normally undertaken by the contractor's own direct employees and those he or she normally sublets; the categories of clients (public, private, and so on) the contractor chooses to work for; and the type of projects in which the contractor specializes (new build, maintenance or refurbishment).

Non-economic factors

Contractor may be selected for reasons unrelated to performance. For instance, the contractor may be appointed to foster business relationship (subsidiary company) or to maintain or promote employment in a locality.

CONTRACTUAL ARRANGEMENT

The client's professional advisers will adopt a suitable contractual arrangement for his or her development. Generally contractual arrangements set out the legal relationship parties wish to establish and hence creates rights, obligations and procedures for resolving contractual disputes. However, in addition, contractual arrangements in the construction industry also establish the basis for making payment to the contractor. The factors which influence the choice of appropriate contractual arrangement include:

1. Size, nature and complexity of development.
2. Dates for commencement and completion.
3. The ability to define the client's requirements clearly before contract.
4. Adequacy of construction information on which to establish client's cost limit.
5. Availability of valid and adequate construction information on which to obtain tenders.
6. The scale of changes the client is likely to effect during the construction phase.
7. State of the national and international economies and their effect on the construction market.

Depending on the client's development requirements, his or her professional advisers may adopt one of the following contractual arrangements. However, the type selected should be incorporated in the tender documentation so that tenderers are aware of the contractual arrangement proposed and their contractual obligations.

LUMP SUM CONTRACTS

Under this contractual arrangement the contractor consents to execute the entire work described or specified for a stated total sum. The agreed sum is based on information derived from drawings, specification, bill of quantities and/or site inspection.

To arrive at the pre-estimated price, the contractor takes into account all contractual risks involved, the condition of the construction market and his or her current workload. The pre-estimated price is paid to the contractor regardless of the actual cost incurred in the execution of the works, providing there are no variations to the works.

The price for lump sum contracts may be either fixed or fluctuating. Financial adjustments to the contract sum reflecting changes in labour and plant rates or material prices during the progress of the works is permissible in the former but not permissible in the latter apart from statutory fluctuations.

Lump sum contracts may be procured either on drawing and specification (for minor works, maintenance, and specialist works) or drawings and the bill of quantities (for all work except minor works).

Lump sum contractual arrangements may be adopted in projects where:

1. Adequate information on which to invite tenders or prepare accurate estimates is available.
2. Fewer variations and strict financial control are expected.
3. The client wants to know his or her total financial commitments before contract.
4. The client has no wish to assume greater proportion of the contractual risk.
5. The client does not intend to retain any control over the details of methods and programming the works.

The advantages and disadvantages of lump sum contracts are summarized as follows.

Advantages

- The client's approximate financial commitment is known before a building contract is initiated.
- The architect is allowed time to develop and hopefully complete the design before contract.
- The project cost can be estimated, planned and controlled during design and construction phases to keep the cost within the client's budget.
- The client can be made aware of his or her cash flow requirements during construction phase.
- The client assumes less financial risks than the contractor.
- The client can initiate some variations he or she can contain during the course of construction.

Disadvantages

- The time taken to complete design before tendering and construction increases the overall project time and cost of finance charges.

○   The client has no control over the contractor's programme and site operations and may influence these only at a cost.
○   Information available for contract documentation may be insufficient and hence may result in the production of inadequate contract documentation.
○   Contractual claims are endemic as a result of late supply of project information and variations ordered by the client during the construction phase.
○   The project may fail to meet time, cost and quality requirements, leading to accusations between designers and contractors.

CONTINUAL CONTRACTS

When the client wishes to pursue a programme of work, his or her professional advisers may recommend one of the following contractual arrangements to effect a saving in tendering cost and expedite production.

Serial contracts

Under this contractual arrangement the contractor undertakes to enter into a series of separate lump sum contracts in accordance with the terms and conditions set out in the initial offer. The standing offer is determined by competitive pricing of a notional bill of quantities containing key items of the proposed projects.

Continuity contracts

Where the client desires to obtain the benefits which arise from continuity of work, the contractor may be asked to enter negotiations based on the original lump sum contract. Once the negotiations are completed and an agreement reached the contractor executes the works as a separate contract within its own parameters.

Term contract

Under this contractual arrangement the client commissions the contractor to undertake specified work within defined cost limits for a definite period, often twelve to twenty-four months. The valuation of the contractor's work is priced on either a schedule of rates or cost reimbursement basis. This type of contractual arrangement is suitable for low value on-going repair and maintenance work where contractors submit invoices for payment on completion of the specified works.

The advantages and disadvantages of continual contracts are summarized as follows.

Advantages

❑   Assured of a programme of works, a contractor can plan ahead and is able to offer the client lower prices in return.

❑   The client saves time and money that would have otherwise been utilized for the preparation of contract documentation for each contract.

❑   Early selection of contractor as well as prompt site operations may be possible.

❑   Long-term relationship enables the contractor to know the client's requirements in terms of quality and speed of execution of the works.

❑   Cost information obtained from previous phases may assist in the provision of improved cost forecasting, planning and control for the subsequent phases.

Disadvantages

○   Subsequent development phases are not awarded on competitive prices and there is no guarantee to suggest that the client is getting value for money.

○   Unsatisfactory performance or delay in one phase may delay the client's entire programme of developments.

○   If terms are not reviewed regularly, the client may pay more for his or her programme of work as a result of a declining construction market that effects lower tender prices.

○   The contractor's business stands to suffer from loss of profit and reduced turnover if the client curtails his or her development plans at short notice.

MEASUREMENT CONTRACTS

Price for units of work under this contractual arrangement is pre-estimated but the total price cannot be ascertained until the work is measured and valued on completion. The evaluation of the measured work is by the application of an agreed unit rate obtained from either bill of quantities or schedule of rates.

Measurement contractual arrangement can be procured on an approximate bill of quantities (when client's requirements are not known in advance) or schedule of rates (when the client's requirements are insufficient to permit the production of bills of approximate quantities).

Measurement contractual arrangements may be adopted in projects where:

1.   The client's requirements are not clearly defined.
2.   There is a need for prompt commencement on site necessitating early contractor selection.

The advantages and disadvantages of measurement contractual arrangements are summarized as follows.

Advantages

❑   Early contractor selection during the design phase may make a prompt start on site possible.
❑   The contractor can contribute to design, advising on quality, buildability, availability of materials, plant and selection of specialist subcontractors and suppliers.
❑   Calculations for interim payments are simplified.

Disadvantages

○   Standard rates when used are often out of date due to infrequent updating.
○   The use of standard rates denies the contractor the opportunity of pricing items to reflect their true cost.
○   The cost of executing specified items of work do not correspond to the rates of comparable items contained in the schedule's standard rates.
○   Contractor stands to lose money if the quantity of an underpriced item increases in the course of the contract.
○   The contractor may have to revise his or her construction programme constantly as the extent of work becomes clear, and this can be the subject of dispute.

COST REIMBURSEMENT CONTRACTS

Under this contractual arrangement the client undertakes to pay the contractor the prime cost: that is, the actual cost of labour, plant and materials utilized in the execution of the works. In addition to the prime cost the contractor is paid an agreed sum to cover establishment charges and profit. This contractual arrangement may be adopted in projects where:

1. The client may wish to influence the execution of the development and hence assumes the entire risk of site operations.
2. An early start is required but the extent of the works cannot be accurately predicted.
3. A high standard of work is required.
4. Emergency, repair and experimental work is needed.

This contractual arrangement may be criticised for lack of financial incentive which would otherwise encourage the contractor to perform efficiently. However, the following variations may be introduced to motivate and enhance the contractor's site performance.

Fixed fee

Under this arrangement the contractor is paid an agreed fixed fee based on the estimated cost of the work. This induces the contractor to work efficiently for profit. However, adequate pre-contract information is required for the preparation of a project estimate on which to agree a fee.

Target cost

Under this method, an estimate is produced for the project and, once agreed, becomes the target price that establishes the basis for the determination of a fee for overhead charges and profit. On commencement of contract the contractor is thereby paid the prime cost and the fee is adjusted to correspond with the increase or decrease of prime cost over the target cost. The main problem with this approach lies in the agreement of a realistic 'target' and the effects of costly variation which takes the construction cost over the agreed target price.

The advantages and disadvantages of cost reimbursement are summarized as follows.

Advantages

❑ Early contractor appointment and commencement of site operations may be possible.
❑ Time saving achieved in the preparation of interim valuations as there is no need for site measurement and percentage assessment from bill of quantities.

❑   The fixed fee approach introduces an element of competition.
❑   High-quality work is assured as the client assumes the financial risks of paying for the contractor's time on site.
❑   Greater flexibility in the issue of variations during the execution of the contract.
❑   Contractual disputes and claims minimized.
❑   Client saves time and tendering cost.

Disadvantages

○   Client assumes higher financial risks.
○   The financial incentive for contractor to work efficiently is minimal.
○   The system is not cost effective and post-contract cost control may be impossible.
○   There is an inherent increased cost of site supervision to ensure that the appropriate working methods are adopted to execute the works economically and that materials and time are not being wasted.
○   The selection of contractors on the basis of lowest percentage addition conceals the fact that a lower on-cost element does not guarantee the contractor's efficiency on site operations.

INTEGRATED MANAGEMENT CONTRACTS

Over the years, several contractual arrangements seeking to integrate the contractor's expertise in both design and construction phases have evolved and under this section the following can be identified for consideration.

Management contract

Under this contractual arrangement, the client appoints an organization or a firm termed as the management contractor to manage and co-ordinate the design and construction phases of the project. This gives the client expertise not normally available at pre-contract stages under the traditional method.

Early selection of the management contractor enables him or her to contribute to the design input. He or she normally provides specified common user and service facilities (e.g. tower crane, scaffolding, site offices, storage facilities, security and so on) but does not execute any part of the permanent works. Rather he or she employs works contractors to undertake the works in subcontract packages. The management contractor is paid for the provision of common user and service facilities in addition to an agreed fee based on percentage of the estimated construction cost for his or her management input.

The system may be adopted for large and complex projects where a contractor's construction expertise is required at the design phase.

The advantages and disadvantages of Management Contract are summarized as follows.

Advantages

- ❑ Early contractor involvement in design may lead to better design and detailing which facilitates construction.
- ❑ The traditional design/construct split is eradicated enabling the contractor to advise on quality, buildability, suitability and availability of labour, plant, materials and construction methods during the design phase.
- ❑ Design and construction is overlapped as well as overlapping the various work packages. This saves the project time.
- ❑ Risks of potential contractual claims is minimized as the managing contractor identifies contentious project information and recommends its modification prior to contract.
- ❑ The 'them and us' attitude is eradicated as the management contractor becomes part of the project team working together to achieve a client's project objectives of time, cost and quality.
- ❑ The criteria for selection of the management contractor is not price alone. In addition, the ability to make some technical and managerial contribution to the design and construction of a project are also considered.
- ❑ Advance purchasing of essential materials and plant can be effected to ensure their availability for use when required.
- ❑ Early contractor appointment enables him or her to give information on the organization of construction works, site layout, possible works contractors and tendering arrangements.

Disadvantages

- ○ Problems of co-ordination between increased numbers of works contractors can lead to delays and grounds for contractual claims.
- ○ Biased contract documentation may be drawn up which unfairly allocates responsibilities to works contractors who may not be well equipped to perform.
- ○ Duplication of site services, attendances and general preliminaries likely.

○ Works contractors, due to financial pressure, may be forced to forgo some of their proper entitlements under the contract.

○ The client suffers financial loss if a works contractor fails to remedy faulty workmanship.

○ The client's financial commitment is not known before the commencement of a project.

○ In the event of the termination of a works contractor's contract due to non-performance, the client bears the cost of arranging the completion contract.

○ The client may have no redress against a management contractor in respect of performance, quality of works contractor's work, late completion and recovery of damages for late completion.

○ Management contractors are (sometimes) pressurized to accept some of the financial risks inherent in construction such as maintaining satisfactory performance and quality of work, time overruns, latent defects and design failures.

○ The client has two sets of fees to pay for the professional services of design and construction management.

Construction management contract

The construction management contract is similar to the contract management approach except that the client assumes a greater share of the financial risks by entering into direct contracts with the various works contractors. The construction manager is a consultant who is employed by the client for a purely managerial role. He or she would not accept any liability for non-completion and so forth, unless resulting from professional negligence.

Project management contract

Under this system the client appoints a construction professional and places on him or her the responsibility of various functions of site identification, land assembly, approval, funding, design, construction and marketing. This construction professional is known as the project manager and, unless he or she is the client's employee, is paid an agreed fee based on a percentage of the project cost. The project manager is responsible for the task of helping to establish the overall project objectives and directing, controlling, planning and co-ordinating the efforts of the project team for the achievement of those objectives. The system is suitable for projects of high value and complexity, short time-scale, larger number of participants, and scarcity of resources.

The advantages and disadvantages of this contractual arrangement are summarized as follows.

Advantages

❏ The project manager performs an independent and disinterested role and uses this position well in directing, co-ordinating and solving construction disputes.

❏ Design can be tailored to overlap construction thereby reducing the overall project time.

❏ The client can be sure that his or her interests are being safe-guarded by the project manager who is an expert in construction matters and in the co-ordination of the inputs of project partici-pants.

Disadvantages

○ The system is not suitable for small simple projects.

○ The individual required within the client's organization to act as a focal point and to oversee that requirements are being met increases the client's construction cost.

○ The project manager requires an executive authority to achieve high performance.

○ The project management carries no financial risks but can be sued for damages if professional negligence or breach of contract can be established.

DESIGN AND BUILD

This method of contractual arrangement enables a firm or an organization to take full responsibility and carry sole liability for both design and construction of a building. After the client has identified his or her need for a building, he or she states the requirement adequately in terms of physical design needs as well as the intended physical use. A selected number of contractors are invited to submit their proposals together with the estimated cost. This method is suitable for standard buildings and industrialized systems (e.g. factories and warehouses).

The advantages and disadvantages of design and build are summarized as follows.

Advantages

❏ The integrated design and construction leads to production efficiency in terms of cost and time.

❏ A simplified contractual arrangement between client and contractor.

❏   Project duration shortened due to contractor's familiarity with his or her system and parallel working on design and construction.

❏   Communication between client and contractor simplified as a result of single link.

❏   The client's total financial commitment known at an early stage.

❏   The client obtains competition in price as well as in design.

❏   The closer contractor/client relationship leads to more efficient design.

❏   The client obtains a design cost element lower than that which an independent designer would charge under other methods.

Disadvantages

○   The contractor's in-house design expertise may be insufficient to solve the client's design needs efficiently.

○   The production of accurate proposals, design and estimate can increase the cost of tendering.

○   An inexperienced client still requires the expertise of professional advisers to prepare briefing documents and tendering information, and to evaluate quality and cost of design.

○   The contractor requires an adequate insurance to cover design failures as he or she assumes the role of design as well as construction; however, this depends on the type of contract adopted.

○   Tender comparison becomes complex as it involves evaluation of design, quality and construction cost.

○   Responsibility for defective design can be complicated by liability dates and time limitations.

○   The client may become stranded with a building unsuitable for his or her needs.

PACKAGE DEALS/TURNKEY

Package deals and turnkey are interchangeable terms for a contractual arrangement in which the contractor offers to provide the client with the service of constructing or providing proprietary systems through press advertisements. Under the system, the contractor may offer to find a site and obtain planning permission prior to price negotiation and construction. The system is employed in the procurement of industrial, commercial and repetitive buildings.

The advantages and disadvantages of package deals/turnkey are summarized as follows.

Advantages

❑ The client saves time in negotiations, starting and completing the project as a result of the contractor's familiarity with his or her system.

❑ The contractor's finished product may be available for inspection by the client.

❑ The client's total financial commitment is known at an early date.

❑ Short line of communication between contractor and client.

❑ Keen competition among contractors enables the client to get a fair deal.

❑ The client enjoys the economies of scale resulting from mass-produced building components.

Disadvantages

○ Variations during the course of construction discouraged as all design decisions are agreed before construction.

○ Control over the quality of work is difficult to achieve in the absence of an independent inspection.

○ Some package designs are not pleasing aesthetically.

○ Limited design available to client.

SEPARATE/DIVIDED CONTRACT

Under this contractual arrangement, the architect carries out the management of the project after the client has approved his or her design solution. He or she arranges on behalf of the client a number of separate contracts with a selected number of works contractors for various work packages. The client enters into direct contract with each of the selected contractors and hence his or her involvement continues beyond the briefing and design phases. The arrangement, which breaks down work into trades or skills, allows the designing and construction of some sections of the works to proceed concurrently. The system may be adopted for most construction projects but is most suitable for a client with an open-ended financial commitment.

The advantages and disadvantages of separate/divided contracts are summarized as follows.

Advantages

❑ Short lines of communication enables the client to make prompt decisions and avoid construction delays.

❑   Reduction of the overall development time by overlapping design and construction phases.
❑   The client obtains prompt utilization of his or her development and earlier return on investment.

Disadvantages

○   The client's commitment on costs is open-ended and he or she would not know the overall financial position until the project is far advanced.
○   The management skills of the client's architect/manager, on whom lies the success of the system, may be lacking.
○   The client must be active in his or her involvement to identify and correct any shortfall and rectify any failings in the management of the project.
○   The system is not suitable for the public sector client who must satisfy public accountability.
○   The client incurs an additional cost in the employment of the services of an experienced site agent to assist the architect in site administration.
○   The client incurs additional cost on contract documentation as he or she invites separate tenders for each work package.

DEVELOP AND CONSTRUCT (THE BPF SYSTEM)

This is a contractual arrangement devised by the British Property Federation as appropriate to their development needs. The system defines the responsibilities of the project participants and rewards them according to performance. Tenders are invited on complete production information on schedule of activities to save time and money. Prime cost and provisional sums are eliminated and the contractor assumes full responsibilities for selecting his or her own specialist subcontractors. To reinforce this point, the successful tenderer is required to sign a document stating that the drawn information is adequate for the execution of the works.

The architect's responsibility is limited to pre-tender design and sanctioning the contractor's detailed drawings. Management of the contract is the responsibility of a hired construction professional or project staff in the client's organization.

The advantages and disadvantages of the BPF system are summarized as follows.

Advantages

☐ The contractor's expertise in detailing quality and buildability are utilized as he or she contributes effectively to the design.

☐ The contractor's detailed design input leads to shorter development time.

☐ The contractor is offered the financial incentive and encouragement to adopt the economies in design detailing and buildability.

☐ The contractor's design input enables contentious design solutions to be negotiated before contract.

☐ As contractors compete on price as well as provision of schedule and programme of activities, resource allocation and method statement, it enables assessment of the contractor's construction and management skills.

☐ The contractor's responsibility for the selection of subcontractors enables him or her to work with firms of his or her choice.

Disadvantages

○ High tender cost incurred, as design and measurement costs are inherent in tenders submitted.

○ The contractor's design input makes apportionment of design liability a complex issue.

○ Design responsibilities forced on the contractor requires his or her early involvement at the design phase but this is seldom implemented.

○ The contractor receives no payment until an activity is completed and this affects his or her cash flow on stretched activities.

○ Whilst the contractor may be competent as a constructor, he or she may not be equally as competent when it comes to dealing with conceptual design solutions.

○ The contractor's choice of materials, and his or her design input and other recommendation, may be biased towards manufacturers who offer high trade discounts.

 RECOMMENDATIONS TO THE CLIENT

As each of the foregoing contractual arrangements has some merits and demerits when adopted, it would be illogical to recommend one to a client without determining first his or her top priority from the following performance requirements:

1. Technical complexity and scale of his or her development.
2. Degree of quality and performance required.
3. Aesthetics/prestige requirements with flexible budget.
4. Economy of construction and tight cost control.
5. Essence of time required and speed of construction.
6. Exceptional size or complexity involving input from many participants.

Each of the above requirements should be rated by the client's professional advisers in order to select a contractual arrangement capable of solving the day-to-day construction operation.

While the above rating may suffice when the criteria for selection of the best contractual arrangement is performance requirement, the following factors also require consideration as they have great influence on the final selection of contractual arrangement best suited to a client's development needs.

Distribution of risk elements

The selection of traditional contractual arrangement, project management or any other option for a client depends on the proportion of the construction risk he or she is prepared to bear. If a client is prepared to accept a greater risk, then the project management option would be the most suitable. But the traditional method would suit a client who wishes to bear minimum risk of the construction.

Type and size of project

If the project is not complex or large in size, then the project management option would be too expensive a method to adopt. Likewise package deal/design build would not suit refurbishment/maintenance contract.

Type of client

While most of the above would be appropriate to clients in the private sector, they will not suit the public sector clients whose construction decisions must in all cases satisfy public accountability. Public sector clients need some competition on price and certainty of financial commitment. Where a client requires to know the price of his or her building and the financial risks he or she is assuming before contract, then it would be prudent to a recommend a fixed lump sum mode of contractual arrangement.

The foregoing suggests that many factors need to be taken into consideration when selecting the contractual arrangement to suit the client's business strategy. There is not a single one which suits all projects; each project has to be treated separately with all the circumstances surrounding it considered. Therefore, the client must be properly informed on the pros and cons of each system. His or her attention should be drawn to the risks: of excessive variation; of delays in releasing construction information; of a decision to accelerate once construction has started; and of starting construction before design is fully complete.

CHAPTER

3

SELECTION OF CONTRACTOR

SELECTION DECISION

Contractor selection for the construction of his or her project is one of the crucial decisions in the client's development ambitions. Ill-conceived decisions on this will mar the client's prospective development gains. The criteria for contractor selection may be price, time or the contractor's expertise in the client's project. If price is the selection criterion, a contractor's tender represents the price which he or she offers to execute the works in accordance with project information shown on the drawings and described in the bill of quantities and contract conditions. This is an objective issue as the client seeks the most economic price for his or her development. However, time and expertise criteria are subjective issues since, because of the necessity to expedite the construction programme and the need for good quality workmanship, the client may seek a price from a sole contractor.

TIMING THE SELECTION

In order to achieve a client's development objectives, his or her professional advisers should, at some stage, decide the most advantageous moment to select a contractor for his or her project. Their concern will be selecting the contractor either before or after design phase. This decision may be influenced by many factors including the following:

1. Type of development – standard system building or a building of specific design.
2. Need for a prompt start on site.

3. Shorter project time requirement.
4. Complexity of development.
5. Certainty of financial commitment before contract.
6. Degree of financial risks the client is willing to assume.
7. Need for strict cost control.
8. Scale of development and number of participants.
9. Availability of resources.
10. A need for early contractor involvement.
11. Degree of unquantifiable risk involved.
12. Business relationships and the client's satisfaction.

For projects characterized by complexity, large scale and shorter completion time, contractor selection during design phase is a better option. However, on projects where the client requires certainty of total financial commitment before contract and strict cost control, the contractor should be selected after completion of design and contract documentation.

 ## MODE OF CONTRACTOR SELECTION

The methods of contractor selection can be described as by competition or by negotiation. In either case, the decision taken should reflect the client's development aims: that is, the completion of his or her development economically and fast, with quality, and at a profit. In a construction project the most effective method of contractor selection is tendering, which may be described as an offer in writing to execute defined work under stated conditions at a price. The types of tendering arrangements available to clients may be grouped under the following headings.

OPEN TENDERING

Under this method, an advertisement is placed in the local, national and technical press by the client's professional advisers, inviting interested contractors to submit tenders for the client's proposed development. In the simplest terms, any contractor who is legally trading may request to tender; however, a deposit is paid for the contract documentation, which is refunded upon receipt of a bona fide tender.

The risk element in this method of tendering to a client is high and this risk is minimized by ensuring that the successful contractor has the required experience and a sound current trading/cash-flow position. However, tender documents are despatched to all contractors who offer to submit tenders for the proposed project.

Public authorities do sometimes invite open tenders and it is a method most suitable for small building and maintenance works and for unusual specialized projects such as asbestos removal and maintenance work to swimming pools.

Advantages and disadvantages of open tendering are summarized as follows.

Advantages

❑ Competitive tenders possible from contractors keen to obtain the work.
❑ It helps contractors willing to grow or expand their market segment to find new clients.
❑ It provides work for local firms who rely on their area public sector client for work.

Disadvantages

○ Escalated cost of tendering in terms of production of increased number of contract documentation and high estimating costs.
○ Less satisfactory contractor performance may leave the client with a bad job.
○ There is no guarantee that all tenders received are bona fide.

SINGLE-STAGE SELECTIVE TENDERING

Under this method tenders are invited either from the current standing list of approved contractors or from a short list of contractors who have responded to an advertisement in the national and technical press setting out details of the proposed development.

The Code of Procedure for Single-Stage Selective Tendering 1991 which sets out guide rules for the system makes useful recommendations, including the following:

1. Depending on the size of the project, five to eight contractors should be invited.
2. Contractors should be given adequate notice of the intended project to enable them to ascertain whether spare capacity exists in their organization to enable them to undertake the proposed work.

3. The contract period should be specified in the tender documents to restrict competition to price alone. However, separate offers based on a revised time are not discouraged.
4. The tender period should be about four weeks. However, longer time may be given for large and complex projects.

Advantages and disadvantages of single-stage selective tendering are summarized as follows.

Advantages

❑   The client achieves an economic price, reflecting what the construction market can bear at the time.
❑   The contractor submits a bid with clear awareness of possible success and spare capacity in his or her organization to accommodate the project.
❑   It satisfies public accountability requirements in contracts for public sector clients.
❑   The client is not responsible for the contractor's site operations, and therefore he or she assumes less financial risks.
❑   Tight cost-control techniques can be applied to ensure that a client's cost limit is not exceeded.
❑   The client can ascertain his or her total financial commitment before contract.
❑   More time is allowed for project design and the client therefore receives better design.
❑   Less abortive tenders and reduction of waste in the construction industry.

Disadvantages

○   The 'short list' principle may exclude suitable contractors.
○   The contractor is excluded from design decisions and his or her expertise is not fully utilized.
○   Time is lost in detailed preparation of the scheme before invitation of tenders.
○   Contractors capitalize on variations issued by the client during the production phase.

TWO-STAGE SELECTIVE TENDERING

Two-stage selective tendering is suitable for clients whose programme requirement does not allow sufficient time to complete design before contractor selection. The circumstances that may bring this situation about include projects where:

1. The benefits to be accrued from early start and shorter construction time exceed the likely risks of commencing the work on half-completed design information.
2. Early contractor involvement to advise on buildability, programming and co-ordination is required.
3. An instant start is necessary before the end of the financial year, leading to saving in taxation or benefits of government grants.
4. Separation of design from construction is impractical: for example, a standard or system building where the contractor can make a considerable technical contribution to the design.
5. Price is only one of the criteria for selection and design input from the contractor is required.
6. The client wishes to assume a greater proportion of the financial risks.
7. Specialist subcontract works form the bulk of the total construction cost and require the input of specialist subcontractors, and their early integration and co-ordination by the main contractor.
8. Extreme shortage of materials and advance purchases of essential materials and components is beneficial to the client.

The two-stage selective tendering system involves competitive selection of contractors in the first phase and negotiation in the second. Competitive selection in the first phase is based on pricing an approximate or notional bill of quantities containing work similar to the project in hand or, when available, pricing the bill of quantities on a similar or related project.

The Code of Procedure for Two-Stage Selective Tendering 1983 recommends a maximum of six tenderers irrespective of the size of project and a minimum tender period of five weeks. The code further recommends the following:

1. Contractors are required to be given adequate notice of the intended project and project information to enable them to decide on their willingness and ability to tender.
2. Acceptance of an additional alternative offer which varies any aspect of the project supported by fully specified and priced build-up.
3. Rejection of tenders in which terms and conditions are qualified.
4. Specified contract period in tender documents to restrict competition to price only.

A contractor is selected after the first stage and is integrated into the design team. As a construction professional the contractor plays an active role in providing advice on quality, buildability, programming and cost. In the second stage, the total contract price is determined partly by negotiation and partly by pricing based on data contained in the first stage. The element of financial risk assumed by the client is greater in the use of this method. The factors that may influence contractor selection under this method include the following:

1. Experience, technical knowledge and ability to execute the work.
2. Capacity, in terms of physical resources (workmanship, plant stores, and so on) and human resources (quality of management, design staff and operatives).
3. Reputation based on past performance on contracts, speed of construction, good quality of workmanship and after-contract service.
4. The contractor's ability to undertake research and development aimed at solving the problems that may be posed by the project.
5. Length of time in business, and a current sound financial and trading position.

The advantages and disadvantages of two-stage selective tendering are summarized as follows.

Advantages

❑ The client benefits from the designer/contractor collaboration during the design phase.
❑ It has the elements of both competition and negotiation in contractor selection.
❑ Early contractor involvement enables him or her to plan for the construction phase more successfully.
❑ An early start on site and early completion can be more easily planned.
❑ Design and construction phases may be overlapped.
❑ Contractual claims minimized as contentious design details modified before contract.
❑ Advance purchasing of essential plant and materials can be planned.

Disadvantages

○ The contractor may be able to alter his or her position/price when the full effect and complexity of the project is known.

○   More effort is required from the client and his or her profes-
     sional advisers than for single-stage selective tenders.
○   Legal responsibilities for design failures are not easily appor-
     tioned.
○   The contractor may impose his or her favourite design solution
     and detailing on the design team.

NEGOTIATED TENDER

Under this method of contractor selection, the client approaches the
contractor of his or her own choice with a view to this being the only
firm who will submit a tender. Contractor selection is based on reputation, specialized skills, and sound financial position, as well as
business relationship.

The process of negotiation can be time-consuming, requiring the
skills and energies of experienced negotiators. To ensure fairness and
to be able to negotiate effectively, the parties to the negotiations
should be of equal or equivalent positions in their respective firms
and possess the same information on which the basis of negotiations
will be established. The negotiation can be based either on the nominated bill of quantities or on cost assessment from a first principle of
estimating.

Nominated bill of quantities approach

Under this method the contractor is obliged to provide a 'nominated
bill'. This is a copy of the priced bill of quantities for a contract that
he or she has won in competition. From the nominated bill the following exercises will be carried out:

1.   An assessment is made of any significant differences between
      the 'nominated bill' contract and the new contract. Onerous con-
      tract conditions, the differences in the tender levels, the state of
      the construction market, and the availability and cost of re-
      sources, labour, plant, materials and finance are all noted.
2.   The contractor is asked to analyse the unit rates of large items
      of work so that his or her establishment charges and profit
      elements are identified.
3.   The contractor will be asked for a breakdown of his or her
      preliminaries in order to establish his or her cost for substantial
      items of plant and site management charges.

Based on this information revised unit rates can be negotiated to reflect all known market factor changes and financial risks that may affect either the client or the contractor in the execution of the proposed project.

First principle of estimating approach

This method is used when a suitable 'nominated bill' cannot be provided due to the magnitude and complexity of the project. In such circumstances the contractor's establishment charges and profit element, and his or her cost on substantial items of plant, site set up and management costs are negotiated.

Once agreement on these factors has been reached, unit rates are calculated for individual items in the bill of quantities from first principles of estimating. This entails utilizing suppliers' prices for materials and components, all-in rates for labour, plant hire rates and domestic subcontractor's prices.

It must be said that during the process of negotiation both parties do not necessarily bind themselves immediately to any of the costs that emerge. The negotiations thus proceed through proposals and counter-proposals of various rates and calculations until finalized. At this point the magnitude of the project is known in terms of cost and financial risks, and the position of eventual agreement is reached, leading to a contract binding on both parties.

Negotiated tenders may be necessary in projects where:

1. The magnitude of the works cannot be assessed before contract (e.g. complex projects).
2. An immediate start is required (e.g. emergency work).
3. The selected contractor can offer a specialized service to a client because of special plant, finance, experience in similar works, awareness of client problems or past satisfactory business associations.
4. The contractor's early involvement is urgently required to assist the design team.
5. The contractor is retained in the execution of a programme of projects.
6. The contractor and the client may have a special business relationship, making negotiation a beneficial option for both their businesses.
7. There are very few contractors with the skill and experience required for the particular type of work.

While the system may be usefully employed in the above situations, it should be recognized that negotiated tenders may be very difficult to agree and may require the efforts of experienced staff from both the client's professional advisers and contractor's offices.

The advantages and disadvantages of negotiated tenders are summarized as follows:

Advantages

- ❑ Early contractor appointment in the design phase leads to a beneficial contribution.
- ❑ Early start on site may be achieved.
- ❑ The client's tendering costs are substantially reduced.
- ❑ The overlapping of design and construction phases condenses the development period with attendant savings in costs.
- ❑ It minimizes tardy delivery of the project as contractor can carry out some construction planning during the negotiation.
- ❑ The client obtains the contractor he or she prefers as the contractor is selected for ability as well as price.
- ❑ The contractor's skills and work experience are made available to the design team.
- ❑ All the important details of the project (e.g. construction programme, method and procedure) are discussed during the negotiation, thus effecting a rational price.

Disadvantages

- ○ The client obtains an offer which is not truly competitive and does not reflect what the construction market can bear.
- ○ There may exist legal implications of joint design.
- ○ Difficulty in estimating on outline information.
- ○ It may not satisfy the requirements of public accountability in projects for public sector clients.

CONCLUSION

Each of the above contractor selection methods has its merits and demerits. No one method is necessarily better or worse than the other and each must be viewed in the light of the circumstances appertaining to the needs of the client.

CHAPTER

4

THE CONTRACTOR'S BUSINESS/ORGANIZATION

 ## THE CONTRACTOR'S BUSINESS NEEDS

The contractor undertakes the execution of unique non-repetitive construction projects within the constraints of resources, quality, cost and time. He or she operates and competes for work with other firms in the unstable markets of the construction industry and his or her source of work is from the clients of the industry.

Like any other business, the contractor's objective in business is financial success and, to achieve this goal and remain in business, he or she adopts business strategies which include the following:

CREATION OF CLIENTS

Client or customer creation is an important objective in the contractor's business as success in this leads to constant flow of business inquiries and/or work. The basis of client creation in the construction industry is: promotion of good public image; reputation in high standard of workmanship; completion of projects on time; avoidance of expensive contractual disputes; business policy towards employees and subcontractors; and creation of new and better construction methods.

PROFIT MAKING

The contractor requires profit to remain in business. He or she needs profit to finance business expansion, and also research and development that will enable him or her to provide the client with improved and varied services. Also, as an insurance against risks in business, the contractor requires a reasonable amount of profit to be able to finance any downturn in business during bad times.

PROVISION OF GOOD PRODUCT

In the construction business, this means providing a good service in terms of high-quality workmanship on time and to the client's satisfaction. To achieve this end, the contractor has to embark on innovation through research and development on building production and management in order to compete successfully with his or her contemporaries in providing high-quality service.

THE CONTRACTOR'S POLICY AND MARKETING STRATEGY

To remain in business, the contractor needs to maintain a constant flow of suitable work at a profit. As the contractor is a price taker – that is, his or her tender price is always in the range the market can take – and generally obtains work through competition, he or she needs a clear defined policy and marketing strategy for his or her business operations. In this light, the contractor needs to consider the following topics:

1. Level of turnover required in the short and medium term.
2. Source of the business finance and how much is required.
3. The type and number of projects which would make-up the expected turnover.
4. The required numbers, quality and specialization of staff and organizational structure needed for the level of operations.
5. Market segment to determine the source of jobs and how these are to be attained.
6. Tender strategy to determine the level of profit mark-ups for the range of projects for which he or she tenders.
7. Construction plans for achievements of high productivity on site.
8. Training, research and development in the field of operation.
9. High standard of workmanship and the build up of a good reputation.

Level of turnover

The level of return on capital is influenced by the number of times capital is employed, that is, the rate of turnover in the course of a given trading period. The rate of turnover indicates how intensively the contractor's capital is being utilized to effect the accumulation process. In the construction industry, the rate of turnover can vary greatly with the size of firms and the nature of their business. It can vary, for example, from 4 to 1 for a small firm which undertakes minor maintenance and repair business to as high as 10 to 1 for large companies dealing in large contracts.

Furthermore, as contractors need to generate enough profit over prime cost to cover minimum overheads, the contractor should strive to attain adequate turnover for generation of sufficient trading surplus to cover his or her overheads. This means that to remain in business, the achievement of a minimum level of output is essential.

Business finance

The amount of finance available to the contractor governs the volume of business he or she can undertake at any point in time. Therefore shortage of finance may preclude him or her from undertaking certain projects and the contractor must ensure that sufficient finance is available for his or her business. As the contractor may not personally raise all the finance required for his or her business, he or she may turn to the following sources for finance:

1. The contractor may supplement his or her own savings by borrowing from friends and relations. This is a cheap source for raising finance but the amount lent is never enough for big business and the borrowed time may be too short. However, it is a good source of finance and many big construction businesses have started or relied on this source of finance in the past.

2. Suppliers of materials operate credit systems which enable contractors to purchase materials on credit. Under this arrangement the contractor can withhold payment of all purchases for a maximum of thirty days. This is a useful facility which enables the contractor to utilize the cash which would have otherwise been spent on materials. However, this facility is lost if the contractor fails to pay the supplier when payment is due.

3. Bankers' loans or overdrafts are the main source of finance for many construction businesses but these are expensive and at times require guarantees and credibility with the banks. In the case of a bank loan the borrower pays the agreed amount of interest on the loan when due. In other words, this is interest due whether or not the loan taken is fully utilized. In the overdraft arrangement, the borrower is allowed an agreed credit limit. He or she can utilize the loan up to his or her credit limit and pays interest only on the amount of the overdraft utilized.

Bankers' loans and overdrafts are reliable sources of finance if one has the required guarantees. At the same time, it puts pressure on the borrower to use the loan effectively.

4.   A building society can provide a mortgage for the business premises or the contractor can raise finance by mortgaging his or her own home.

5.   A contractor can take on a partner to raise additional finance for the business. This is a good source of finance, but again the amount required may not be enough to expand the business as required.

6.   A contractor may invite people to take up shares. For a private company this option is open only to family and close relations. However, the limited liability company may raise finance by floating all forms of shares for subscription by public and corporate bodies. In an ordinary share issue, the shareholders are paid a proportion of the company's profits. They receive return on their investment only when the company makes a profit and the extent of that return depends on the magnitude of profit. The return for debenture and preference shares does not depend on the profitability of the company. A fixed return is payable when it becomes due regardless of the company's trading result. This makes debenture and preference shares risky and expensive sources of finance.

The contractor should decide which of the above methods of finance suits his or her business operations and act accordingly.

Type and number of projects

The type and number of projects the contractor undertakes at any point in time depends on many factors including the following:

1.   *Contractor's business finance*   As was mentioned above, the contractor's operating capacity is limited by his or her financial standing. In all construction contracts, the contractor is required to finance the project until such time as interim payments by the client makes the project self-financing several weeks or months later. If the contract duration is long, the cost of funding to the contractor can be enormous. Therefore the size of the available finance will dictate the level of the contractor's business activities and the type of projects he or she undertakes, be it multimillion pound contracts or jobbing work.

2.   *Work and management experience*   In the construction business it is beneficial for the contractor to select projects in which he or

she is experienced and capable of doing well. Project inexperience leads to serious mistakes and faulty workmanship that can be expensive to rectify. If the contractor is experienced in building and refurbishing houses, this experience is not the same as building new hospitals or banks. Although it can be said that management expertise can always be acquired by recruiting experienced staff on the appropriate employment markets, staff on 'one-off jobs', however, may become redundant when the project is completed.

3.   *Jobs available in area of operation*   The contractor's choice of jobs is limited by the type of jobs available in his or her area of operation. If jobs in the area of the contractor's operation are refurbishment of offices, factories, warehouses, conversions of residential properties into offices, and so on, the contractor should not look for or engage in newbuild projects.

4.   *Number and quality of subcontractors*   The variety of trades necessary to construct projects are too diverse for the contractor to master. There is, then, the need for most trades to be executed by other firms; this leads to reliance on the expertise of subcontractors. Therefore the contractor would only undertake projects if he or she knows that the services of good subcontractors can be obtained in the locality of the project.

5.   *Spare capacity in the contractor's organization*   The contractor should match the type and size of projects undertaken with the spare capacity in his or her organization at all times. This requires constant monitoring of projects in hand, rate of progress and finishing times in order to assess the level of spare capacity and the type and size of projects for which to bid. Failure to do this may lead to problems when the contractor discovers that he or she has taken on projects for which he or she has no spare capacity to execute efficiently. This certainly leads to delays and possible payment of damages for non-completion.

6.   *Organization*   Organization is the creation of a formal structure for the attainment of the contractor's business objectives. It plays a key role in determining the success and profitability of any business. Organization brings into existence division of functions, authority centres and human interactions. The size, style and structure will be determined by the contractor's style of management and line of business, and the company's annual turnover.

Organization structure

To be successful, the contractor should have a thorough knowledge of the capabilities of his or her organization to ensure its effectiveness in undertaking construction projects. Only staff with the appropriate skills and experience should be recruited and the number of staff should equate to workload. All bottlenecks must be removed and there must be a free flow of accurate and relevant information at all times. Figure 4.1 shows the organization structure of a typical medium/large-size construction company.

The following is a résumé of the functions of the various departments/sections in the organizational diagram shown in Figure 4.1.

1. *Drawing office*   This office provides drawings for design and build contracts as well as sketch details to supplement the architect's working drawings.
2. *Planning and site supervision*   This section is under the control of the contracts manager and is responsible for planning, preparation of programme charts and provision of the site liaison.
3. *General office*   The general office keeps records, provides secretarial and administrative support services for the technical and operating departments.
4. *Costing*   This department monitors the cost of all labour, materials and plant used on all projects undertaken. This information enables the contractor to reconcile the cost with the value of a project at any given time.
5. *Accounts*   This department is in charge of keeping records of all monies received (for works executed or sold) and expended on salaries, wages, purchase of goods, assets and liabilities. It also prepares the contractor's annual trading account.
6. *Wages*   Payment and keeping records of wages (including all the necessary deductions for income tax, national insurance and so on) of staff and operatives are dealt with by this department.
7. *Plant*   This department keeps and maintains all the contractor's plant and ensures that all plant is in good working order and is available when required.
8. *Jobbing and small works*   This department undertakes small alterations, repairs and redecorating work.
9. *Surveying*   This department administers contract accounts, which are composed of preparation of interim valuations, final accounts, contractual claims, cost reports, valuation of variations, and calculation of bonus payments.
10. *Buying*   The buying department keeps records of various suppliers and their current prices and terms of supply; obtains

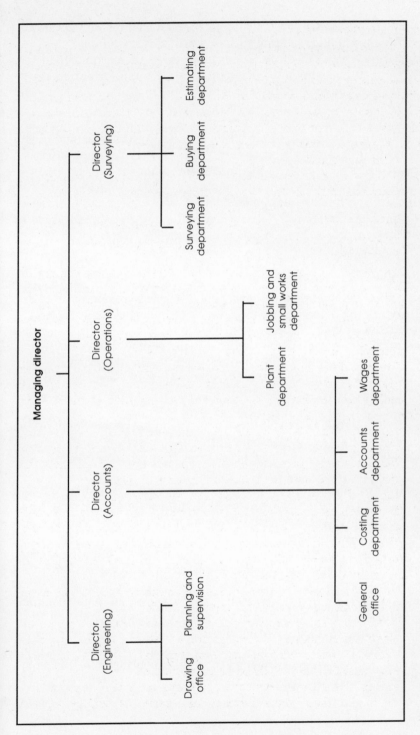

Figure 4.1 Organizational structure of typical medium/large size construction company

quotations from various suppliers; and measures and orders the required materials from the cheapest source of supply.

11.   *Estimating*   This department undertakes the preparation of estimates and provides management with the information necessary for converting the estimate to tender.

All the above sections or departments together form the contractor's business organization. Although they do not directly take part in site operations, their respective inputs are necessary for the successful completion of all projects the contractor undertakes. The costs of running these sections or departments (e.g. wages, salaries and directors' fees; rents; office equipment and services; depreciation and so on) are known collectively as the contractor's establishment costs or head office overheads.

Establishing market segment

The contractor should state clearly his or her area of operations. This will ensure the avoidance of the expense of undertaking work located beyond his or her market segment. However, as the construction market is not static or stable, the contractor may be compelled, at times, to look further afield for work when the local market dries up or becomes uncompetitive. The contractor normally seeks business in the following ways:

1.   Replying to advertisements in the technical press. Some public sector clients (e.g. the local authorities) place an advertisement in the technical and local press when they require the services of contractors.
2.   The contractor's reputation, or by making his or her abilities and expertise known to clients or professional advisers, he or she may be selected either to compete for a job or to enter into negotiated contract.
3.   Placing an advertisement in the local press stating area of work specialization as in speculative design and build or system building.
4.   A satisfied client may retain the contractor for a development programme.
5.   Personal contact with architects, engineers and surveyors in private practice; influential people in business and politics; officers in public service such as in the local authorities, and so on, may lead to an important source of work.

Maintenance of reliable client

The selection and maintenance of reliable clients should be the contractor's prime target. A reliable client is one who does not disrupt the contractor's progress on site, makes prompt interim and final account payments, and settles and pays contractual claims promptly.

Accurate estimates and successful tenders

The contractor must first put in an accurate bid and, secondly, must be successful before he or she can think of making a profit. In an effort to put in reasonably accurate bids and be successful in tenders the contractor should be aware of the cost of his or her resources. These constitute the elements of cost (plant, labour, materials and establishment charges) used in the build-up of the unit rates. Each of these is treated fully in Chapter 5 (pp. 73–9) and need not be exemplified here.

Monitoring site operations

Accurate successful tenders still require the appropriate input from site staff/operatives, and the effective control of resources and cost, in order to sustain a profit. Winning the contract on accurate bidding alone, without an effective site organization, cost control, cost feedback from site operations and the appropriate paperwork, leads to loss of profit. Therefore there is the need to plan the site efficiently for it to make financial contribution to the contractor's business. A high degree of site control, costs analysis and cost control must be constantly exercised to ensure the success of the project.

1. *Site control*   The exercise of effective site control ensures proper allocation of well-defined tasks, setting targets and ensuring that the targets are not lost by reason of:
    (a)   poor performance;
    (b)   use of unplanned or unspecified method; or
    (c)   use of untrained or incompetent labour.
2. *Cost analysis*   This is a financial exercise which aims at the conversion of the total physical events on site to financial terms:
    (a)   to ascertain the total financial effect of various site events;
    (b)   to identify the main areas of excess cost; and
    (c)   to determine the variances of costs of events from financial allowance in the bill of quantities.
3. *Cost control*   This is a continuous cost exercise which compares, in financial terms, what is happening on site in relation to what was budgeted to happen. This exercise is intended:

(a)   to allow positive action to achieve the original plan; and

(b)   to allow a logical basis for financial review.

The cost control exercise would comprise the reconciliation of the cost of executing items of work with the value of expected receipts from interim payments; if there is a shortfall in the expected receipts, then the following action must be taken to identify the operations creating loss in order to institute an immediate corrective action:

(a)   The effectiveness of the site management and operation must be reviewed. If this proves that there is laxity in the operation, corrective action must be taken immediately.

(b)   If the site operation proves to be on course, the next check should be on the accuracy of quantities in the bill of quantities. There may be undermeasured items that have caused the shortfall in payment. If this should be the case, remeasurement and valuation may rectify the payment situation.

(c)   If (a) and (b) prove satisfactory and accurate, then the rates which won the contractor the job may be low and, if so, there is a need for improved estimating and tendering procedures.

Reputation and quality of service

The contractor's reputation and the quality of service he or she provides are important factors that influence contractor selection in both competitive and negotiated tenders. Therefore it is vital that the contractor considers the following factors:

1.   *Completion of project on time*   This enables the client to gain early use of his or her project or an early return on investment. It is also beneficial to the contractor financially if he or she completes projects on time to avoid the payment of damages for non-completion. The contractor's reputation is enhanced if he or she is listed for the efficiency of construction management, for planning and providing for problems before they arise, and for successful early completion of projects.

2.   *High standard of workmanship*   Although high standard of work starts from the drawing board, the contractor plays an important role in maintaining quality of the highest standard on construction production. He or she should ensure at all times that his or her material and workmanship, and those of his or her subcontractors, are not below the standard specified. He or she

should also follow all the good construction practices which lead to product of a high quality.

3. *Prompt attention* Quick response to the client's requests or complaints demonstrates a professional attitude, which always promotes good business relationships. The contractor should ensure that clients receive prompt attention at all times in their development and maintenance needs.

4. *Good after contract service* The client often requires the contractor to make some minor changes, additions or improvements to the project after commissioning. It is important that the contractor gives attention to these requests, however small in cost. Failure to respond to a call to provide maintenance service to a building he or she has constructed suggests that the contractor is avoiding his or her construction mistakes and this perception can be damaging to the contractor's business.

5. *Good technical advice* The contractor has a wealth of knowledge on construction matters and when this expertise is required by the client or the design team during design or construction, sound technical advice should be given at all times to effect economy and time. Poor advice may lead to bad and costly decisions affecting the contractor's reputation.

6. *Training, and research and development* For training staff to execute effectively the work they are employed to do is essential to the contractor's efficient operations. At the same time, conduct of research along the lines of the contractor's area of business might result in the discovery of better working methods or new techniques leading to improved services to clients. Although both staff training and research cost money and can only be undertaken by firms with adequate resources, the results can be rewarding, thus making the expenditure worthwhile. The contractor should therefore decide on the percentage of his or her business returns to be allocated to staff training, and to research and development.

7. *Public relations* The creation of a good public image is important to the contractor's success in business. In this light, the contractor should plan how well he or she presents his or her company publicly in terms of press advertisement, glossy literature, proper location of office and signboards, protecting the public from dangerous substances, safe boundary fences with viewing panels on tidy sites, maintenance and better treatment of staff, subcontractors, and so forth. All these points may seem unnecessary expenditure, but the rewards from a good public image can be enormous.

RESOURCE MANAGEMENT

In order to remain in business and operate competitively and profit-ably, the contractor's attention should be directed towards an efficient use of his or her resources. This means making sure that those re-sources – capital, manpower, plant and materials – are effectively managed and utilized to achieve competitive and successful tenders that lead to profitability and growth.

EFFICIENCY OF CAPITAL

Be it his or her own or borrowed, capital has an element of cost. The contractor loses interest that would have been earned on own capital invested elsewhere and, at the same time, borrowed capital will incur interest charges. Therefore capital should be utilized effectively to finance itself and generate profit needed for business survival. To embark on this strategy the contractor must ensure that, capital being value in motion, it should always be in motion to retain its character as capital. Capital is said to be effectively utilized when money capital, productive capital and labour power are efficiently employed to produce commodity capital that is quickly transformed to money capital to produce surplus value. Funds locked up in business transactions do not contribute to the production of surplus value and therefore cease to be capital. Speed and efficiency of this cycle of transformation are crucial to the contractor's business; thus any slow down in the motion is not helpful. The contractor must ensure that his or her finance is released quickly in all projects undertaken. In these circumstances he or she should consider the following factors:

1. Embarkation on projects that ensure free motion of capital. Projects which require financing by the contractor or with infre-quent flow of payments should be approached with caution.
2. The contractor must ensure no undervaluation for interim pay-ments. Valuations must include all works properly executed to date, all variations on architect's instructions executed, and claims for loss and expense, in order to ensure a healthy cash flow.
3. Projects must become self-financing at the earliest possible time.
4. Agreement and settlement of final accounts and contractual claims must be prompt in order to release money locked up in disputes.

5.   The contractor must ensure that projects are completed on time and must avoid all unnecessary delays whenever possible.
6.   Excessive wastage of materials during their use on site must be avoided.
7.   Expensive specialist staff should be hired only as needed when not required or affordable on a full-time basis.
8.   Reliance on hired plant and tools, and purchasing only those items of plant for which continued and constant use can be guaranteed.

COST OF CAPITAL

Cost of capital is considered under the following headings:

Payments to shareholders

This consists of interest and dividend payment offered to money capitalists who are debentures, preference and ordinary shareholders in order to attract investors to the money market. The extent of this cost to the contractor depends on his or her choice of finance and the risks, rewards and securities he or she offers against the borrowed funds. Lenders of unsecured loans and ordinary shareholders will demand that they receive higher rewards in recognition of the higher risks they have assumed. Debenture and preference shareholders and those lending money on secured loans may not make such a demand if the level of risks they assume is lower.

Payments to financial institutions

This comprises interest payments to banks, building societies and other financial institutions who granted the contractor some money capital for the business. Here again, the magnitude of this interest cost will be determined by the security the contractor offers and the risks assumed by the lenders.

Payments for financial advice and services

Administrative cost involved in raising the capital arises from the services of financial experts when the contractor chooses to raise capital on the open market by share flotation.

EFFICIENCY OF LABOUR

Directly employed labour in the contractor's main and site offices must work efficiently for the survival of the contractor's business. These employees are responsible for what is to be done, what payment to expect, where funds are coming from, how the payment is made and who receives it on the contractor's behalf. In effect, they are the backbone of the contractor's business and they must be motivated to perform efficiently.

The contractor's site managers and operatives are the next important category of workers on whom the success of the contractor's business lies. Besides skill, the contractor should realize that the efficiency of the site workforce depends on weather conditions, work organization, effectiveness of plant and tools employed, and combination of operatives, some of which are beyond their control. Therefore this calls for greater managerial efficiency in the organization of site operations.

Increasing productivity

Increased competition for a share of the construction market requires higher levels of productivity. The contractor needs to employ work study and operational research in the establishment of productivity standards and better working patterns. These techniques may be summarized as follows:

1. *Work study* The work study technique is composed of method study and work measurement. Method study seeks to improve production methods by the adoption and application of:
   (a) better site layout and construction methods;
   (b) effective equipment and handling techniques; and
   (c) better working conditions;
   and work measurement seeks to establish the basis of comparison and control to facilitate effective programming by undertaking:
   (d) assessment of performance of labour and plant; and
   (e) application of costs to expected performance.
2. *Operational research* This technique seeks to match methods and time for the production of better work systems.
3. *Other motivating factors* While the above techniques may induce a better working method and improved conditions, operatives may be unwilling to perform efficiently. The site may be plagued by strikes, work to rule, go slows and absenteeism. All these factors affect productivity and it is in the interest of the contractor to try to reduce the disruptive effect of his or her

operatives. In addition to the techniques mentioned above, the following may be adopted to effect productivity:

(a) Maintenance of the health of operatives by the adoption of safe site and working practices (e.g. provision of better protective clothing, good lighting and thermal comfort, and prevention of exposure to severe weather conditions and excessive noise).

(b) Maintenance of psychological factors such as better working relationships, co-operation, job satisfaction, motivation, and the selection of healthy and able operatives.

(c) Operatives remunerated adequately by the adoption of better reward schemes: financial (bonus and profit sharing), and semi- and non-financial (good pension, social clubs, subsidized catering facilities and promotion status, job security).

COST OF LABOUR

The cost of labour is made up of wages and emoluments paid to operatives and statutory costs incurred in the employment of labour. The individual elements of cost of labour is fully discussed in Chapter 5 (pp. 73–4).

EFFICIENCY OF PLANT

The employment of plant in construction work reduces the amount of physical energy required in the execution of some operations. It also accelerates the pace of the construction process by facilitating the execution of some mundane work of lifting heavy materials and transporting. The contractor therefore requires the use of plant in the execution of some sections of the project. It is therefore essential that he or she keeps a record of all up-to-date plant for construction work.

Plant utilization

Plant hire or ownership is expensive and the contractor should ensure its effective utilization once acquired. Economic utilization of plant depends on:

1. Sufficiency of work available to keep the plant in maximum working capacity during the period of acquisition.
2. Experienced and skilled plant operator.

3. Availability of adequate supply of reasonably priced power.
4. Matching capacities of all plants and avoidance of under-utilization.
5. Plant usage being cheaper when compared against manual labour.
6. Site layout and conditions facilitating the economic use of plant.
7. Plant selected being suitable for the terrain of the work.
8. Easy access to maintenance and repair facilities in case of breakdown.
9. Adequate storage area while on site or in contractor's yard.
10. Careful planning and continuous programming of work required to reduce wasteful idle standing time.

Hiring and purchasing plant

The small contractor undertaking jobbing work and whose meagre capital is fully committed may rely solely on hired plant. But as the business grows and the work load increases he or she may decide on outright purchase of specific items of plant for the business. However, he or she will need to consider availability of capital outlay, state of current and expected workload, length of time the plant may be required, expected rate of utilization and proper storage and maintenance facilities before embarking on any purchases.

If the contractor can afford the initial and running cost involved and can predict the utilization factor, then he or she may decide on plant ownership. Nevertheless, it may prove advantageous for the contractor if he or she hires specialized items of plant (e.g. excavators, tower crane, scaffolding) for short periods and own the general types of plant (e.g. concrete mixers, dumpers, trucks).

The advantages and disadvantages of plant hire are summarized as follows.

Advantages

❏ Contractor's capital is not locked up in expensive pieces of equipment and so cash flow is not affected.
❏ Hiring cost can be paid monthly as the contractor is paid for work done in interim payments.
❏ The contractor does not pay for plant he or she does not require or has no work for.
❏ Modern and suitable plant is readily available on hire from plant hire companies at fixed hourly, daily, weekly and monthly rates.

❑   The cost of maintenance repairs and replacement is borne by the plant hire company.

❑   In many instances the plant hire company provides experienced and skilled operators and the contractor does not have permanent operators' wages to pay.

❑   The contractor has neither storage problems nor cost of maintaining expensive plant departments.

Disadvantages

○   Effective planning and programming of work schedules is required to minimize expensive idle time on hired plant.

○   A suitable or required machine may not always be available when needed.

○   The contractor still has to pay for the hire cost when work is aborted due to bad weather.

○   The plant operator does not work for the contractor full-time and so there is no incentive to give his or her best.

○   Plant may not be in good shape to undertake the work; constant breakdown will cost the contractor dearly in delays.

○   Hire rates depend on market forces, which fluctuate and can be a cause of loss to the contractor if the hire cost escalates.

Also the contractor loses the following facilities when he or she hires an item of plant in lieu of outright purchase:

○   Ownership of a purchased item of plant immediately transfers to the acquiring company, which as a company asset can be used as a security to raise more finance if required.

○   Outright purchase of an item of plant entitles the acquiring company to capital allowances – a good source of tax saving.

○   Credit sale (an arrangement whereby title to the item of machinery transfers on purchase but the acquiring company works off the purchase price by instalments) – which is another good source of tax savings if it can be negotiated.

Purchase of plant

Total cost of own plant is made up of initial cost (capital cost plus interest charges on capital over the period of advance of capital) and running cost (insurance, spares, supervision, testing, maintenance, and so forth).

While the contractor can be certain of the initial cost, by reason of the fact that fixed interest charges can be arranged for a short period, he or she cannot be so sure of the extent of running costs, which depend on the quantity of the work the plant will do, amount of maintenance the plant will require and rate of deterioration. The cost of the plant will be high if it is underutilized during its economic life. While frequent breakdown increases maintenance cost, plant rate of deterioration may be accelerated if it is overused without adequate maintenance. However, the rate of obsolescence depends on the rate of technological progress which is outside the control of the contractor. The contractor must ensure that plant is economically utilized and does not fall into disuse before the end of its useful economic life.

In addition to the uncertain costs cited above, the contractor incurs the following additional costs in utilizing plant be it owned or hired:

1. *Cost of transport* This is the cost the contractor incurs when an item of plant is transported from plant depot to site and vice versa. The extent of this cost depends on the distance from depot to the site.
2. *Cost of working base* When, for example, tower cranes are to be employed in the site operations they require a suitable concrete base or tracks to work on. This is a cost which the contractor incurs.
3. *Cost of erection and dismantling* The contractor incurs the cost of having items of plant (e.g. hoists, scaffolding, tower cranes) erected and dismantled.
4. *Cost of fuel and power supply* The contractor requires fuel and power supply to run the plants on site and should pay for their supply.
5. *Cost of operators* The cost of machine operators and their mates are costs to be considered by the contractor.

The above costs should be carefully considered when the contractor plans for the use of an item of plant in his or her operations. Whether hired or owned these costs remain unchanged and should be allowed for in the contractor's tender submissions.

MATERIALS

Materials used in the production process are not produced by the contractor and so must be obtained from elsewhere, that is, direct from either material manufacturers or builders' merchants. Whatever the source of the supply the contractor should ensure that the materials purchased are:

1. Of good quality, and purchased at the right price, in the correct quantity, and from a good and reliable source of supply.
2. Delivered to the right site at the appropriate time, and are carefully unloaded, stored and adequately protected.
3. Incorporated into the building with the right constructional methods to avoid wastage.

Neglect in any of the above factors means an unnecessary expenditure on materials and a shortfall in the expected profit margin.

Good material

The material specified in the bill of quantities, specification or on the drawing is the right material required for the project. As any other material is not acceptable, the contractor should exercise great care in ordering the type specified. To comply with the specification, the contractor should re-quote the specification, all British Standard references, and any other relevant guidelines, when ordering. Where the contractor is not sure on these points, the architect should be consulted and all discrepancies cleared up before purchasing.

Quantity

To ensure that the right quantity of materials are purchased, the contractor's buyers should derive the required quantities of materials by measuring from the architect's drawings. Apart from the main objective of establishing the right quantities of materials, this exercise serves as a check against the amount contained in the bill of quantities.

The quantity of materials purchased should include a factor of waste normally paid for by the client. The contractor should note that while overpurchasing will cost him or her money in loss of profit, underpurchasing and omissions from the materials schedule means delays in the construction programme and the risks of damages for delayed completion.

Source of supply

Unless the bill of quantities specifies the source of supply, the supply source of all materials is at the contractor's discretion. The key indicator to the best source of supply is the supplier's terms as this has a great influence on the material price. It pays, therefore, to examine the benefits of the following terms-of-supply range prior to purchase:

1. Cash purchase only or cash on delivery.
2. Applicability of cash discount on cash purchase only.
3. Thirty days' credit facilities.
4. Purchase price including or excluding delivery and unloading charges.

Purchase price

Right price determines how competitive that price is when all the terms of supply have been considered. The right price should include the selling price, and the terms and date of supply. A cheap material with late supply date is of no use to the contractor as it will lead to delay in project completion and, in consequence, the possibility of damages.

Delivery to the right site

For the contractor with only one site, the right site is not a problem. However, where the contractor has many sites, off-loading urgently needed material at a wrong site is a possibility. To prevent this happening the contractor should ensure that the correct site address accompanies all orders and, where possible, suppliers are provided with directional information.

Time for delivery

Right time is the time when the materials are required for the building operation. Materials which require site assembly or conversion should arrive on site earlier to enable specified work to take place prior to incorporation into the building.

The delivery time must be compared with the cost of storage. Premature purchasing should be avoided as storage is costly in terms of provision of storage space, cash-flow drain, and damage to and loss of materials stored on site. However, where there is likelihood of material shortage in the future, then advance purchasing is better than the situation where work is at a standstill from lack of materials.

Unloading and storage facilities

Unloading of materials requires care and attention in the selection of the right unloading plant, which, if not owned, can be expensive to hire. The storage facilities should also be adequate for protecting the materials from the weather, damage and theft.

Method of incorporation

Materials should be incorporated into the building carefully applying the best working practices. Care should be exercised to avoid misuse of materials (e.g. the application of materials in wrong places, the wrong proportion of materials in mixing, and substitution of materials).

Waste factor

Conversion waste is an unavoidable waste which arises during the use of the material. This waste factor is allowed for in the build-up of unit rates for work items contained in the bills of quantities. However, this allowance should not be exceeded during purchasing or incorporation into the works.

Material reconciliation

The contractor should carry out this exercise at the end of the project to reconcile total materials purchased, and those returned or resold, with the actual quantities incorporated into the work. This information will help the contractor in his or her future material purchasing, management and planning.

 # SUBCONTRACTING

As construction is a complex process and no one can master all the trades it entails, it requires the input of many specialist firms in the production of buildings. These specialist firms who work with the contractor in the building production are known as 'subcontractors' to whom the contractor sublets the execution of part of the works.

EFFECTS OF SUBCONTRACTING IN THE CONSTRUCTION INDUSTRY

The subcontracting arrangement has reached the stage where main contractors, in most instances, sublet all the work and retain only the managing and financing functions. The contractor, therefore, enjoys freedom from maintaining directly employed operatives and plant; this has resulted in a disinvestment in the construction industry. Contractors are no longer responsible for the training of operatives whom they do not employ directly; also, they neither invest in plant and machinery nor undertake any research and development. These important issues are left in the hands of subcontractors who are sometimes without the resources to cope. Operating with only office and few specialist staff, the contractor of today is able to switch his or her capital from the construction business to other forms of investment (e.g. the stock market), when the activities in the construction market are down and vice versa.

In subcontracting, the contractor relies on the subcontractor's prices when preparing tenders for projects (this will be explained in detail in Chapter 5).

TENDER ACTION

 ## TENDER INQUIRY

Once a contractor has achieved a good reputation on previous projects he or she receives tender inquiries from prospective clients from time to time. The contractor should promptly consider each inquiry in compliance with the requirements of the Code of Procedure for Single-Stage Selective Tendering 1991 and let the prospective client and his or her professional advisers know whether or not he or she will tender for the project.

 ## PRELIMINARY INQUIRY

A preliminary inquiry to a contractor should provide him or her with all the relevant information regarding the project and its principal participants. The following information may prove useful to the contractor.

The name of the client

Awareness of who the prospective client is enables the contractor to ascertain whether he or she is a new or an existing client. If an existing client, the contractor will know how he or she gets on with that client in business. However, if it is a new client, the contractor will try to obtain background information on his or her reputation, financial standing, and the way in which he or she conducts their business.

Information on the client's professional advisers

The names of the client's professional advisers (e.g. architect, quantity surveyor, structural engineer, services engineer, project manager) enables the contractor to establish whether he or she has had any previous dealings with them and how they conduct their business generally.

Description and value of project

With this information the contractor is able to verify the scope of the works and determine if he or she is capable of managing such a project successfully with his or her current resources.

Location of project

The knowledge of the location will enable the contractor to assess whether the project is within his or her operational area. If it is not, he or she will have to ascertain the availability of materials, labour, subcontractors, and so on in the locality.

Form of contract

Familiarity with the form of contract assists the contractor in decision making. An unfamiliar form of contract or contract conditions may prompt the contractor to decline the offer. However, the contractor may decide to conduct research and gain some appreciation in the form of contract proposed.

Tender action period

The proposed dates for dispatch of tender documents and tendering period assists the contractor in assessing if he or she can take on an additional estimating commitment within the stated time. It also helps him or her in planning the estimating workload.

Contract period

The information on the likely project commencement date and contract period enables the contractor to know if he or she can take on an additional project. This information also helps him or her to plan job allocation for his or her current resources. However, in some cases, time is not a factor of competition and hence not stated in the inquiry document.

DECISION TO TENDER

The contractor's decision whether or not to tender is taken after identification and evaluation of all the possible risks relating to the proposed project. The following factors will assist him or her in his or her response to the initial tender inquiry.

Construction and estimating workload

The contractor's current workload may be heavy and may remain so for many months ahead with no spare capacity to take on more. Alternatively, current projects may be nearing completion and he or she may therefore have spare capacity for new commitments before long.

State of the construction market

The construction market may be in a state of decline due to economic/political conditions and the contractor cannot afford to turn down inquiries, and vice versa.

Level of turnover

The contractor may be experiencing a shortfall in his or her targeted annual turnover. In that state, he or she will have to redress the shortfall by responding positively to inquiries.

Contractor's business strategy

The contractor may plan to expand his or her business and therefore requires new clients, some of whom may be outside his or her operational area. On the other hand, strategically, he or she may plan for contraction in business and therefore deliberately decline inquiries.

Client and his or her professional advisers

The contractor should consider the following points:

1. *Client* Does the client have either a sound financial standing or backing for the project?
2. *Professional advisers* Has the contractor dealt with them before and can therefore anticipate the standard of finish required? If not, what facts can he or she discover about their expecta-

tions, the way they run projects, and the manner in which they conduct their business generally?

Characteristics of project

The contractor should consider the following issues regarding the proposed project:

1. *Type*   Is the project within the category of work the contractor usually undertakes efficiently and profitably?
2. *Size*   Is the size of the project within the range which the contractor's current resources can cope with?
3. *Competition*   The level of competition for this type of work may be high (e.g. a prestigious project) therefore making it not worthwhile to tender.
4. *Programmed time*   Can the project be commenced in the stated time, so as to slot into the company's current and projected workload?
5. *Form of contract*   Is the contractor familiar with the proposed form of contract and any special conditions incorporated?

Competitive advantage

The contractor will consider the level of competitive advantage he or she may have over the firms likely to compete for the project.

Availability of resources

The contractor should consider the following issues regarding the availability of resources:

1. *Materials*   Materials for that type of project may be in short supply and alternatives may only be obtained from abroad, with the possibility of causing uncontrollable construction delays.
2. *Plant*   Availability of specific plant items and their maintenance for that type of project.
3. *Labour*   Are suitable labour resources and subcontractor available in the area of the proposed project?

TENDER DOCUMENTS

If the contractor accepts the invitation to tender, complete tender documents will be sent to him or her for the preparation of the bid. The tender documents normally comprise the following items.

Bill of quantities

This may be a document with firm or approximate quantities and may be subdivided into trades, elements, operations or activities.

Drawings

These will consist of general arrangement drawings showing site, floor and roof.

Specification

This document will apply where a bill of quantities is not in use. This document may consist of descriptions of quality of materials and workmanship with an indication of the amount of work required.

Form of tender

This is the document containing the tender sum and the contractor's signature that is submitted. A bill of quantities is not submitted with this form unless it is a specific requirement.

Letter of invitation

This is a covering letter that may contain various items such as date of return of tender, expected starting dates, project peculiarities and all relevant tender information.

Return envelope

This is an addressed envelope for the return of tender to ensure that each tenderer has the same and correct address of the client.

SEQUENCE OF TENDER PREPARATION

To ensure that no part of the estimating process is overlooked, the use of ordered and logical methods of estimating is essential. In order to achieve conciseness in the preparation of the estimate the contractor may adopt estimating sequences summarized as follows.

INSPECTION AND INTERPRETATION OF TENDER DOCUMENTS

On receipt of the tender documents the contractor's estimator makes a brief inspection of the documents. He or she records his or her findings on an Enquiry Record Form which he or she forwards to management for decision whether or not to tender. During the inspection the estimator checks the following items.

Affirmation of information

The estimator checks that the information provided in the initial inquiry on which the decision to tender was based has not altered. In the case of alteration, the financial effect of the change has to be assessed and noted on the Enquiry Record Form. For example, a change in commencement date for the project may affect the contractor's operating programme; while a change to a higher value or an unfamiliar form of contract may cause the contractor to decline to tender.

Technical complexities of project

The estimator will consider if the contractor has the resources to cope with any technical complexities and challenges which the project poses. If the contractor's current resources are inadequate a case may be made for hiring more to undertake the project.

Adequacy of project information

Inadequate project information does not facilitate a smooth and successful contract. The estimator will therefore make certain that the project information is adequate, and free from abnormalities and discrepancies.

Time allowed for preparation of the tender

An adequate time reflecting the type of project and its size should be allowed for the preparation of the tender. Adequate time allowance facilitates a thorough appreciation of the project and the collection of all relevant information, leading in turn to reasonably accurate estimates.

Form of contract

The estimator will check the form of contract stated. Where a familiar form of contract is stipulated, the legal and financial implications of any amendments will be noted by the estimator.

Information on physical conditions of site

The physical characteristics of the site (e.g. access, layout, circulation and storage space, type of ground, water table, dangerous or unpleasant conditions) influence factors such as construction method, plant selection and productivity. The clarity of this information is important for the preparation of an accurate estimate.

Restrictions imposed by the client

Where restrictions have been imposed by the client (e.g. phased construction, reduction of noise and working hours, presence of client's tradesmen), these should be clearly stated in the tender document. The estimator will assess the legal and financial implications of these restrictions.

Time allowed for construction

Time allowed for the construction depends on the size and complexity of the project, and where this is considered insufficient the estimator will make a note of this and recommend an increase in tender sum to cover the cost of possible overrun on the contract programme.

Value of contractor's input

The estimator will compare the proportion of the project value to be undertaken by the contractor with that to be undertaken by specialist subcontractors. If the value of the former is smaller than the latter, it indicates a project with a high services element. Should that be the case, then the contractor is expected to manage and co-ordinate the input of a large number of specialist subcontractors on site. However, with different levels of 'specialist' subcontractors, most work is now subcontracted.

Quality of design

The estimator will examine the quality of the tender documents. Any traces of haste in their preparation will be noted by the estimator as it is an indication of extensive variations during construction.

Method of measurement

Deviations from a familiar method of measurement and their financial effect will be recorded by the estimator.

Method of payment

The estimator will record the method of payment proposed (e.g. periods of payment, and percentage to be retained by the client). If the value of retention is high it may mean a strain on the contractor's cash flow and higher interest charges on borrowed funds to finance the project.

Damages for delay

The estimator will also check the magnitude of liquidated and ascertained damages imposed by the client for delay, and where this is considered to be high he or she may recommend an increase in the tender sum to cover this risk.

Management's decision whether or not to tender will be influenced by the information the estimator provides on the Enquiry Record Form as a result of his or her study. If management decides to tender for the project, the estimator will proceed with procurement of all estimating information and prepare an estimate for the project. However, if management decides otherwise, the tender documents should be returned, accompanied by a letter giving polite reasons for declining the offer.

GATHERING ESTIMATING INFORMATION

As the tender period is generally limited, once the decision to tender has been made, it is essential that the estimator determines and programmes key dates for certain information and activities in advance. For example, the latest date for dispatch of enquiries, receipt of suppliers' and subcontractors' quotations, and visits to the site and the offices of the client's professional advisers are all essential items to be programmed. Other examples of important dates are those for in-house meetings to discuss construction methods and agree a method statement and construction programme.

DETAILED EXAMINATION OF TENDER DOCUMENTS

As the tender documents contain most of the estimating information, the estimator's next priority in the estimating process is to examine them thoroughly. Likewise, it is advisable for the construction team (contract manager, planner, project surveyor and the buyer) to study the tender document and to have an appreciation of the project. This knowledge enables them to contribute effectively to the estimating process when required.

BILL OF QUANTITIES OR SPECIFICATION

During the thorough examination of the bill of quantities or specification, the key points the estimator will note include the following:

1. Unfamiliar and contentious clauses for discussion at the adjudication meeting.
2. Discrepancies between the documents (especially between the drawing and the bill of quantities/specification) for clarification with the client's professional advisers.
3. Sections of work the contractor intends to sublet, work normally executed by specialist subcontractors, and key materials and components to be purchased from specialist suppliers.
4. Adequacy of descriptions and their effect on the unit rate build-up.
5. Definitions of the contractual relationship between the contractor and his or her specialist subcontractors and suppliers, particularly when the main contractor has to supply materials for a subcontractor who will be on a labour-only basis.

CONTRACT DRAWINGS

General arrangement drawings normally accompany tender documents. Detail drawings may be inspected at the architect's office by appointment. The estimator should make the following observations in the examination of the detail drawings:

1. The quality of the project development and design.
2. The complexity or simplicity of the design.
3. Whether any of the details will require special plant or skill in its execution.
4. Whether operations are repetitive (e.g. standard size of concrete sections allowing repetitive use of formwork).
5. Whether the quality of finish, standard of workmanship, degree of accuracy and the tolerances required can be achieved with the existing site managers and operatives.

While in the architect's office, the estimator should sketch and make notes of matters that might affect and alter the construction methods envisaged and also study the site investigation report and make notes or photocopies of important sections. In addition he or she should raise all dubious points with the architect for clarification. He or she should note particularly which designs are incomplete and are likely to be carried out after tender.

PHYSICAL CONDITIONS OF SITE

Information on physical conditions likely to be encountered can only be obtained on a site visit. The estimator normally visits the site in the company of a construction manager, site agent and plant manager. While on site the estimator should prepare a site visit report covering many points, including the following:

1. Location of the site in relation to existing infrastructure (e.g. roads, rail, electricity, gas, water and sewerage).
2. Site layout and topographical details and their effect on operations, storage and accessibility.
3. Ground conditions and distance to tip. In particular, the possibilities of making substantial savings in the disposal of surplus excavated material.
4. The level of security required.
5. Availability, likely skill and experience of local labour.
6. Police regulations regarding traffic restrictions.

Table 5.1 Method statement of college library project

Description of items	Quantity	Details of method	Plant	Output per week	Plant labour involved	Period report
Excavate pipe trench	500 m	Excavate backfill Plant	Backacter	500 m	4 labourers	1 week
PVC pipes	500 m	Lower by hand	Nil	250 m	6 labourers	2 weeks
Basement excavation	4 000 m^3	Excavate direct Load to lorry	Backacter and lorry	2 000 m^3	2 labourers	2 weeks
Basement reinforcement	5 000 kg	Supplied cut and bent	Nil	1 666 kg	2 steel fixers	3 weeks
Basement concrete	400 m^3	Site mixed	14/10 mixer	100 m^3	6 concretors	4 weeks

CONSTRUCTION METHOD APPRAISAL

The estimator requires information which may include historical output and cost, method appraisal and availability of resources. Hence he or she should meet with the personnel who would be responsible for managing the project and providing plant items and other services when the tender is successful. At the meeting, alternative methods of construction will be considered and evaluated, taking into account time and current resources at the contractor's disposal. A method statement will be prepared from the selected option on which the estimate will be based. An example of a method statement is given in Table 5.1.

TENDER PROGRAMME

Having a thorough appreciation of the information contained in the tender documents and having decided on the construction method to adopt, and the sources of plant, labour and materials, a pre-tender programme is drawn up. The programme is a comprehensive statement of the contractor's intentions with regard to sequence, timing and duration of operations (see Figure 5.1). This document is important to contractors as it may be quickly adapted and presented as a master programme.

A pre-tender programme would indicate production patterns, gang structure, likely seasonal changes and interdependency of trades or operations. This enables the estimator to build up a full picture of their likely effects on these items of costs and provide adequately for them in the unit rates build-up.

SUBCONTRACTOR'S PRICE ENQUIRIES

At site level, the contractor's suppliers and domestic subcontractors play important roles, and therefore before the calculation of the unit rates has commenced, the estimator sends out price enquiries to these organizations. These enquiries are, in effect, an invitation to a limited number of suppliers and subcontractors to submit competitive quotes for the supply of materials or for the execution of items of work contained in the contract documentation. To ensure consistency and fairness in approach, avoidance of omissions or misunderstanding of important points, and compliance with the requirements of the code of procedure for Single-Stage Selective Tendering 1991, the estimator should incorporate the following in the inquiry document:

Month	June				July				August					Sept	
Week ending	6	13	20	27	4	11	18	25	1	8	15	22	29	5	12
Week number	8	9	10	11	12	13	14	15	16	17	18	19	20	21	22
Item of work															
RC foundations	▓	▓	▓	▓	▓	▓	▓								
PC frame								▓	▓	▓	▓				
Concrete sundries												▓			
Blockwork cladding												▓	▓		
Refuse chutes													▓		
Block partitions														▓	▓

Figure 5.1 Programme chart, Contract No. 703: college library project – section programme

1. A photocopy of the relevant sections of the original tender documents.
2. Description of project, its location and approximate value.
3. An indication of the magnitude of a particular subcontractor's or supplier's involvement.
4. A statement of an approximate date for commencement of the project and when the subcontractor's input is required, the form of contract in use, and any contractual restrictions.
5. Information on site facilities and services offered by the contractor (e.g. use of plant, scaffolding, messroom and responsibilities for unloading, storage and protection).
6. Terms of payment and damages for delay clearly spelled out.
7. A date for the return of quotations, to ensure that all quotations are returned on a set date for evaluation.

On return, the enquiries should be examined and evaluated. In the case of a material supplier's quotation, the following examination should be carried out:

1. Conformity of quoted material with specification.
2. Availability of material for the specified construction period.
3. Suitability of delivery arrangement.
4. Level of trade and cash discount offered.
5. Period the quotation remains open for acceptance.

For domestic subcontractor's quotations, the following checks should be carried out:

1. Consistency and competitiveness of rates.
2. Level of discounts offered.
3. Comparability of terms of quotation with those in the main contract, particularly inflationary fluctuations.

The level of pricing is not the only criterion for selecting a subcontractor. In addition to price levels, the estimator should consider if the following responsibilities have been clearly defined:

1. Unloading, storage, protection of materials and components, and moving same to work positions.
2. Protecting the completed subcontract works until handover.
3. Provision of special plant and equipment required for the subcontract works.
4. Provision of site facilities such as office space, messroom and water.

5. Design and consequences of design failures (where work requires subcontractor design input).
6. Acquisition of specific materials and components for the subcontract works.
7. Planning and scheduling the subcontract works to meet the contract programme.
8. Financial stability of the subcontractor enabling him or her to undertake work and meet construction programme.

The most competitive quote to emerge from the above scrutiny will be considered (after the necessary adjustments) in the project estimate. However, in practice, main contractors invariably negotiate significant discounts at a later stage in order to obtain the best possible tender levels with subcontractors.

All tenderers should be notified promptly of the results of the enquiry to enable the successful subcontractors and suppliers to plan for action in case the main contractor's tender is successful.

 ELEMENTAL ANALYSIS OF COST

Price is the total remuneration the contractor receives from the client for executing an item of work. This amount, in effect, is the client's cost and comprises the contractor's total cost plus profit.

The contractor's total costs are made up of sums of the following individual elements of cost: labour; mechanical plant; materials; and overhead or establishment charges. In the preparation of the estimated cost of the project, the estimator should establish an 'all-in rate' for each of these elements of cost and use it in building up unit rates for each measured work contained in the bill of quantities.

LABOUR

Labour cost

This comprises wages and statutory employment costs. The gross labour rate for an operative is therefore composed of basic wages for a 39-hour minimum working week fixed by the National Joint Council for the Building Industry, and incidental costs composed of the following:

1. National Working Rule allowance.
2. Guaranteed time (e.g. wet weather).
3. Guaranteed minimum bonus.
4. Non-productive overtime.
5. Productivity payments.
6. Sick-pay schemes.
7. Training and CITB levy.
8. National Insurance.
9. Pension contributions.
10. Holidays-with-pay schemes.
11. Severance pay.

Gross and net hourly rates

The sum total of basic wages for labour and the incidental costs listed above is known as the 'gross hourly rate' for labour, and if used in calculating unit rates for various measured works contained in the bill of quantities, this is termed as *all-in pricing*. The alternative is known as the 'net hourly rate' for labour where only the basic wages for labour (without the incidental costs) is used in computing unit rates for the measured works. If this pricing method is adopted, it is termed *net pricing* and in that case, all incidental costs are each priced separately in the preliminaries section of the bill of quantities.

MECHANICAL PLANT

As the use of mechanical plant is predominant in most construction work, it is important that the cost of plant is properly represented in the cost of items of work. A realistic approach is the determination of the required items of plant and the assessment of the length of time they will be required on site. The length of time plant is required is related to the construction programme which indicates the amount of work the plant is expected to do.

The type of plant selected and its planned usage influences allocation of its costs in the estimates. Where plant serves a number of trades or operations (e.g. a hoist or tower crane), an allowance in the preliminaries or project overheads would be the appropriate cost allocation. But where plant is to be used for specific operations (e.g. excavator or concrete mixer), its operating cost may be allocated to unit rates for the items of work on which it is employed.

The all-in hourly cost information for items of mechanical plant may be obtained from plant hire firms, published schedules of plant hire charges, the in-house plant department, or calculation from first principles (see Example 5.1).

Calculation of plant hourly rate

In calculating all-in hourly rates for plant from first principles, most items of mechanical plant are given a working life of five years, however dumpers have a working life of two years and concrete mixers three years. Plant generally does not work 100 per cent of its life due to incidence of breakdown, repairs, idle time, and severe weather conditions, and in essence plant may be taken as working only 75 per cent of its lifetime. Example 5.1 gives an example of the hourly rate build-up.

Example 5.1 Build-up of hourly rate

Assume capital cost of plant £15 000 with an estimated working life of 5 years, working 1 800 hours per annum; scrap value £2 500; interest charges on borrowed capital $12^1/2$% per annum.

		£
Capital cost	=	15 000.00
Interest at $12^1/2$% per annum	=	9 375.00
Maintenance and repairs (allow) 10% per annum on capital	=	7 500.00
Insurance (allow) £300 per annum	=	1 500.00
		33 375.00
Less Scrap value	=	2 500.00
÷ 5 years		30 875.00
÷ 1 800 hours per annum		6 175.00
Hire rate per hour		**£3.43**

COST OF MATERIALS

The cost of unit material specified for a section of the work comprises the following items:

1. *Purchase price*   This is the price quoted by the supplier and paid by the contractor less any trade discounts.
2. *Transporting*   Transport cost is incurred where delivery charges do not form part of the price paid for the material and the contractor collects the material from the supplier's premises.
3. *Unloading*   Where this does not form part of the sale agreement the contractor incurs an additional cost of unloading. This operation can be expensive where bulky materials such as precast concrete units and structural steel are involved. There is also the cost of inspecting and checking material for right quality and quantity during this operation.
4. *Storage*   Storage space for fragile and expensive materials such as glass products and ironmongery to prevent damage and theft incurs expenditure to provide.
5. *Handling*   This is the cost of transporting material from storage to its place of use. This can be a costly operation if it involves long haulage distances.
6. *Conversion*   This concerns the cost of any work done to the material prior to its incorporation into the works. For example: converting cement, sand and aggregates into concrete; cement and sand into mortar; timber into windows, doors and frames.
7. *Return of empty carriage*   This cost is borne by the contractor and this again can be expensive where the distance between site and the supplier's premises is long.
8. *Waste*   Although the cost of direct or unavoidable waste is borne by the client, in practice the contractor finds it difficult to stick to this waste allowance. Indirect or avoidable waste occurs to the material during storage handling, conversion and incorporating phases of the construction process (see pp. 90–3).

For each material being considered, the total of the above items of cost may be divided by the quantity of materials required to establish unit costs.

OVERHEADS

Each of the elements of cost (labour, plant and material) can be related to a specific item of site production. Any element of cost which is not confined to an individual item of work may be conveniently described as a project overhead. Project overheads cover financial matters that relate to the entire project and may be considered under two subheadings.

Site overheads

These cover the cost of providing site management and equipment required for the execution of a project. This cost is not incurred unless a project is undertaken and hence is known as direct cost, and it includes the cost of maintaining site supervisory staff, offices, temporary roads, security, mess facilities, and power. The tender documents normally include provision for these general project items. In estimating this cost the estimator takes into account the contract period and provision, erection, maintenance, and clearing cost.

Head office overheads

This element of costs is sometimes described as an establishment charge and includes all costs the contractor incurs in the running of the business. Like the site overheads, it cannot be related directly to any individual item of work. It is also not confined to a particular project and the expense is incurred once the contractor is in business and has no bearing on whether or not he or she undertakes contracts. Head office overheads may include cost items shown in Example 5.2.

The magnitude of head office overheads is influenced by the following factors.

Annual turnover

This is the expression of the contractor's intended value of business in a given period (usually one year). The annual turnover is established by the contractor's management and, once this has been done, a percentage of the expected return on the turnover is allocated to finance the cost of establishment charges. This element of cost is normally expressed as a percentage and allowed in the estimates in one of the following ways:

1. In the case of net rate pricing, the percentage is used in calculating a lump sum for inclusion in the estimate (see Examples 7.2 and 7.3).
2. In the case of gross rate pricing, the percentage is added to the net unit rates for each measured work contained in the bill of quantities (see Example 7.4).

Example 5.2   The build-up of head office overheads

	£
Builder's overheads*	
Interest on capital	8 500
Interest on bank loans (allow)	5 000
Rent of offices	6 250
Rates	2 000
Salaries or wages of staff *not* directly connected with the contract	70 000
Cost of head office telephone, heating and lighting	1 800
Fees	4 000
Stationery, postage, sundries	3 500
Company pension	10 000
Insurance for offices	1 500
Depreciation on head office equipment and small plant not costed elsewhere	2 000
Directors' fees and expenses	10 000
Transport not directly concerned with particular contracts, including running costs	8 000
General expenses	1 450
Total cost of running firm per annum	£134 000

Builder's overhead percentage

With assumed turnover of work of £1 775 000 per annum the percentage overhead will be $\frac{£134\ 000}{1\ 775\ 000} \times 100 = 7.55\%$

*Note: This is not an exhaustive list of all possible overheads, but covers the usual main headings.

Type of work undertaken

The contractor's rate of turnover may be influenced by the type of work he or she undertakes. If he or she undertakes contracts where a large amount of working capital is tied up in materials, plant, contractual delays, irregular payments and claims, his or her turnover rate will not be sufficient to generate the required return to pay for the overheads.

Staff abilities and efficiency

If office staff are efficient there will be no overmanning in any section of the organization and hence less overhead costs. To keep staff overhead costs down, the contractor should recruit, train and retain the right calibre of staff, and promote efficiency.

Composition of overhead

The factors contributing to head office overhead cost elements differ in each firm. Therefore, the estimator needs to be clear about the individual cost elements which compose this total cost in order to make the appropriate allowance for it in the estimates.

 WINTER WORKING

Where the construction period is over one calendar year, sections of the work will be executed during the winter months. The estimator should therefore consider factors of cost associated with winter working and allow for them in the estimate. Factors of cost normally considered may include the following:

1. *Protective shelters for operatives*   For example, partial protection by polythene panels or sheets secured to scaffolding for mild winters, while full encapsulation by polythene-covered timber-framed panels secured to scaffolding, or supported structures (and even tents) for low-level work, should be considered for severe winter working.
2. *Protection of materials*   For example, bricks, concrete blocks, steel reinforcement, lintels, timber, and drainage goods should be protected by steel-framed structures covered with polythene.
3. *Protection of new work*   For example, new concrete work and brickwork require this protection up to seven days after comple-

tion and may involve complete encapsulation in polythene sheets, with some insulation and heating introduced.

4. *Loss of production*   For example, operatives confined only to their protected work areas with limited space to manoeuvre seldom produce efficiently.

5. *Extra material or equipment*   For example, steam jet heaters – heating coarse aggregates, water, formwork and insulations – are required to achieve the correct setting and hardening times of concrete; also, air entraining agents that allow for freezing expansion are added to mortar for brickwork.

6. *Need for artificial lighting*   For example, portable electric lights or floodlights secured to poles or scaffolding are an ideal proposition.

COMPILING THE ESTIMATE

After collecting all the relevant facts and information, the estimator commences the process of calculating the total estimated cost of the project. A logical approach is required in this process so that omissions and errors are averted. The following sequence may be recommended.

PREAMBLES

The preamble clauses should be studied to ensure that the price sought is for the quality of materials specified. Reasonably accurate estimates should be provided for all workmanship clauses carrying large monetary value to compensate for time consumed in preparation or high wastage factor in meeting the requirement. Unfamiliar workmanship clauses should be brought to the attention of management at the adjudication meeting.

BUILDER'S OWN WORK

Traditionally, the work undertaken by the builder's own operatives varies, but excavation and earthworks, concrete work, masonry, woodwork, drainage and fencing are work sections builders usually keep in-house. It must be mentioned that as most sections of work are now

subcontracted out, this procedure is rapidly changing. Never
in estimating the cost of work to be executed by the builde
workforce, the estimator builds up unit rates from first pr......
using an all-in rate for labour, plant and material.

SUBLET WORK – DOMESTIC SUBCONTRACTORS

If the contractor intends to sublet any sections of the work, he or she
is often required to submit a list of proposed domestic subcontractors
with the tender. The sections of work contractors usually sublet vary
but may include demolition, asphalt work, roofing, metalwork, plumbing, glazing and surface finishes. To price for these items of work, the
estimator adjusts the most competitive quotation obtained from domestic subcontractors by adding a percentage to cover overhead, profit,
attendance and elements of risk.

SPECIALIST SUBCONTRACTOR/SUPPLIER WORK

Sections of work specifically intended to be executed by specialist
firms are covered by prime cost sum in the bill of quantities. Prime
cost (PC) sum is defined as various sums of money written into the
bill of quantities either for works which are required to be carried out
by a nominated subcontractor or goods and materials required to be
obtained from a nominated supplier. To enable the client to receive or
accept a tender that represents the total cost of all the works, the
contractor is obliged to include these sums in his or her tender and
hence is given the opportunity to price for profit and attendance.
Specialist work varies but may include piling, structural steelwork,
electrical and mechanical installations, and patent glazing.

Attendance on specialist subcontractors

The attendance contractors provide to specialist subcontractors may
be classified either as general or special. General attendance means
granting a specialist subcontractor the use of contractor's facilities
such as standing scaffolding, messroom, sanitary accommodation and
welfare facilities, space for office accommodation and for storage of
plant and materials, light and water.

Special attendance, on the other hand, refers to attendance provided to a specialist subcontractor over and above the general attendance provisions and this includes cost items such as access road,
hardstanding, storage, power, unloading, distributing, hoisting and

placing in position. These items have to be stated in the bill of quantities (see Example 5.3).

Example 5.3 Examples of typical prime cost sums

(a) PC sum for specialist subcontractor's work

Include the following PC sum for work to be executed by a nominated specialist subcontractor:	£	p
Electrical installations including estate lighting and television aerial system:	10 000	00
Add for profit %		
Add for general attendance		
Add for special attendance including unloading materials and plant, storing, distributing, hoisting or lowering, providing power supply to maximum load of 5 kW.		

(b) PC sum for materials to be supplied by a specialist supplier

Include the following PC sum for goods to be supplied by a nominated specialist supplier:	£	p
Ironmongery	500	00
Add for profit %		

WORK COVERED BY PROVISIONAL SUM

A provisional sum is a sum provided in the bill of quantities for work or for costs which cannot be defined or detailed due to uncompleted design.

In order to eradicate the confusion and the disagreement that the inclusion of provisional sums in the bill of quantities creates, it is the requirement of the SMM7 (Standard Method of Measurement, 7th edition) that all provisional sums be classified as either undefined or defined. The reasons for this classification may be summarized as follows.

Undefined provisional sums

The contractor is not expected to make any allowance in his or her contract programme, plan or price in the preliminaries for items of work covered by undefined provisional sums. Therefore the contractor is reimbursed from the provisional sum for any extra costs he or she incurs on these items (see Example 5.4).

Example 5.4 An example of undefined provisional sums		
Provisional sums for undefined work as SMM general rules 10.5 and 10.6	£	p
Contingencies	13 500	00
Metal windows	300	00
Additional works to adjoining owner's boundary wall and fences	3 500	00

Defined provisional sums

The contractor is required to make an allowance in his or her contract programme, plan and price in the preliminaries for items of work covered by defined provisional sums (see Example 5.5). Therefore no claims for reimbursement will be entertained. To enable the contractor to provide adequately for items of work described under the defined provisional sum, the following information regarding the work must be made available to him or her:

1. The nature, extent, quantity and quality of the work.
2. The location of work, how it is executed, limitations and whether subsequently covered.

SPOT ITEMS

Spot items represent alteration work to an existing building and hence are required to be priced on site. On his or her site visit the estimator analyses the work content of spot items, measures where necessary and assesses a lump sum cost including overhead and profit for inclusion in the estimate (see Example 6.1).

Example 5.5 An example of defined provisional sums

Provisional sums for defined work as SMM general rules 10.3 and 10.4	£	p
Insurance in accordance with Clause 6.2.4 of the Conditions of Contract	500	00
Building Control Officer's fees	4 000	00
Work to internal courtyard; comprising reduced level excavation, pea shingle filling, precast concrete pavement boulders, topsoil and planting; approximate area 30 m^2	3 000	00

It is important to note that where provisional sums are given for defined work, or work by local authorities, named subcontractors or statutory authorities, the contractor will be deemed to have made allowance in the preliminaries/general conditions for all preliminaries costs in connection with such work.

DAYWORK

During the progress of the works, any varied work which, by its very nature, cannot be properly measured and valued, is calculated at daywork rates based on prices prevailing at the time of work. The normal practice is that a provisional sum (probably undefined) is included in the bill of quantities to cover any dayworks. The intention of such provision is to provide a standard basis of evaluating work which can only be assessed on a prime cost basis. Therefore, at the tender stage, the estimator is asked to insert a percentage addition required on labour, material and plant to cover overheads and profit together with other costs not included within the prime costs (e.g. weather uncertainties, bonus).

Classification of daywork

As the stage of a project at which daywork is carried out influences supervision and overhead costs, daywork is classified as follows:

1. Daywork carried out during the contract period.

2. Daywork carried out after the date of commencement of the defects liability period.
3. Daywork in connection with any work which does not form part of the main contract.

Prime cost of daywork

The prime cost of daywork comprises the following items:

1. *Labour charges* Labour charges consist of standard wage rates and expenses for the standard working hours. This is computed by multiplying the wage rate by the time spent directly on the daywork. An example of the calculation of a standard wage rate is given in Example 5.6.
2. *Material charges* These consist of the prime cost of materials and goods from either suppliers or manufacturers including the cost of delivery but excluding trade discounts, or contractor's stock valued at the current prices plus any appropriate handling charges.
3. *Plant charges* Plant charges are the cost of those items of plant and transport engaged in carrying out the dayworks. The rates used in calculating the daywork are those contained in the current schedule of plant charges issued by the Royal Institution of Chartered Surveyors (RICS).
4. *Overhead and profit* As stated above, competing contractors are normally asked to state the percentage addition for daywork they require to cover their overhead costs and profit as part of their tender. As contractors have diverse operating costs and business strategies, this percentage addition will vary from firm to firm according to:
   (a) type and size of project;
   (b) location in relation to head office;
   (c) contractual risk;
   (d) state of order book;
   (e) state of the construction market;
   (f) level of competitive advantage;
   (g) geographical area; and
   (j) construction period and rate of construction output.
   Therefore it is an item which is considered after enough knowledge of the project has been acquired. In addition, the following list of incidental costs should be studied and their respective cost implications assessed:
   (a) cost of site staff and site supervision;
   (b) additional cost of overtime;

Example 5.6   An example of the calculation of a standard wage rate

	Hours	Hours
Standard working hours per annum		
52 weeks @ 39 hours/week		2 028
Less 21 days' annual holiday as follows:		
16 days @ 8 hours/day	128	
5 days @ 7 hours/day	35	
Less 8 days' public holiday as follows:		
7 days @ 8 hours/day	56	
1 day @ 7 hours/day	7	226
		1 802

Calculation of hourly rate

	Craftsman £	Labourer £
Standard basic rate	160.00	136.00
Guaranteed minimum weekly earnings		
47.8 weeks/annum	x 47.80	x 47.80
	7 648.00	6 500.80
Employer's National Insurance contribution		
(allow) 9%	688.32	585.07
	8 336.32	7 085.07
CITB annual levy @ 0.25%	20.84	17.71
Public holidays (included in guaranteed minimum weekly earnings above)	–	–
Annual labour cost	8 357.16	7 103.58
÷ 1 802 hours		
Hourly base rate	£4.64	£3.94

Example 5.7 Summary of daywork

	£	p
Work executed in accordance with SMM7, Clause A55		
Labour		
Include the undefined provisional sum of £ 5 000.00 for labour.	5 000	00
Add for incidental costs, overheads and profit. %		
Materials		
Include the undefined provisional sum of £1 000.00 for materials.	1 000	00
Add for incidental costs, overheads and profit. %		
Plant		
Include the undefined provisional sum of £500.00 for plant in accordance with *Schedule of Basic Plant Charges January 1981 issued for use in connection with Daywork under a Building Contract* as published by the Royal Institution of Chartered Surveyors (RICS).	500	00
Add for incidental costs, overheads and profit. %		

Specialist trade work executed in accordance with SMM7, Clause A55

Specialist trade	**Incidental costs, overheads and profit**
Labour	_____ %
Materials	_____ %
Plant	_____ %

Specialist trade	**Incidental costs, overheads and profit**
Labour	_____ %
Materials	_____ %
Plant	_____ %

(c) time lost due to inclement weather;
(d) extra bonuses and incentive payment;
(e) apprentice study time;
(f) extra employer's national insurance contribution;
(g) sick pay;
(h) subsistence and periodic allowances;
(j) fares and travelling allowance;
(k) third party and employer's liability insurance;
(l) tool allowance;
(m) protective clothing, safety and welfare facilities;
(n) non-mechanized plant usage (e.g. standing scaffolding, staging); and
(p) cost of use, repair and sharpening small tools.

After the above study and evaluation, the contractor's decision on the level of percentage addition for dayworks will depend on his or her assessment of contractual risks, tendering policy and commercial judgement. However, the following percentages are normally included in tenders by contractors:

(a) labour plus 100–200 per cent;
(b) material plus 10–30 per cent; and
(c) plant plus 5–25 per cent.

Example 5.7 shows the provision of daywork in the bill of quantities.

ESTIMATING

GENERAL POINTS

DEFINITION

Estimating is a technical function undertaken to assess and predict the total cost of executing an item of work in a given time using all available project information and resources. This exercise involves the synthesis of the elements of individual cost of materials, labour and plant to arrive at a net unit rate. The net unit rate plus an allowance for profit and overhead is the gross unit rate which the estimator places against a measured item of work contained in a bill of quantities when the gross rate pricing method is adopted (see p. 266).

LABOUR CONSTANT

Labour constant refers to a standard time, that is, time taken by a skilled operative to complete a task in a controlled environment. As site operations do not take place in a controlled environment, labour constants cannot be usefully utilized in estimating without some adjustments. The estimator therefore relies on the historical performance of the contractor's operatives on past projects, work study information and experience to effect the necessary adjustment and establish reasonable labour time for the various items of work contained in the bill of quantities. In this respect, it must be appreciated that the labour constants given in this text are for reference only.

RELEVANCE OF DEEMED INCLUDED ITEMS OF SMM7

For simplicity and brevity of measured work items, the SMM7 states that some items of minor works (which could be separately measured) should be considered as part of relative work items. Hence these minor work items are not mentioned in the bills of quantities and it is incumbent on the estimator to identify these items of work and their cost implications and include them in the unit rate build-up.

ERRORS IN THE BILL OF QUANTITIES

All errors or discrepancies between the tender documents discovered by the estimator should be brought immediately to the attention of the client's quantity surveyor who will attend to these matters and notify all tenderers of any corrections to be made in tender documents. To expedite the process, tenderers may be informed of these corrections first by telephone, and then have them confirmed later in writing (see Figure 6.1).

On receipt of notice of amendments, the estimator should make all necessary corrections and return the signed copy of the amendment to the client's quantity surveyor.

WASTAGE OF BUILDING MATERIALS

The following two types of material waste are encountered in the building production process.

Avoidable waste

This is an incidence of material waste which the contractor can avoid by good material management policy, and through training site and head office personnel. There are, however, other participants in the construction process (e.g. the architect and the quantity surveyor) who contribute to this waste but who are outside the contractor's organization. This type of waste can be further categorized as follows:

1. *Direct waste*   This type of waste involves either complete loss of materials during the production process or damage of materials beyond repair. Direct waste of materials can occur during material transportation and delivery to site, site storage, internal transport to works, cutting installation, and learning by operatives and/or apprentices.

Super Qess Partnership
Chartered Quantity Surveyors
1A Upper High Road
London
EC1 4CM

Our Ref:
Your Ref:
Date:

*For the attention of Mr Smith
(Chief Estimator)*

Dear Sirs

Proposed Erection of College Library – Amendment Number 1

Further to my letter inviting tenders for the above scheme and our telephone conversation, will you please amend your copies of the bill of quantities as follows:

Bill Reference		*Heading*
Page 2/96	L 404	Delete 'rough cast and clear float'
Page 3/48	Item S	Delete 'Clause F323' *Insert* 'Clause F309'
Page 3/57	Items A-G	Delete 'Ref 3'
Page 3/64	Item D	Delete 'Clause L 406' *Insert* 'Clause L 404'
Page 3/64	Item F	Delete 'Clause L 436' *Insert* 'Clause L 423'
Page 3/67	Item A	Delete 'Clause M 100' *Insert* 'Clause M 108'
Page 3/67	Item D	Delete 'Clause M 259' *Insert* 'Clause M 405'
Page 3/111	Item C	*Insert* '22 nominal size' before 'in floor screed'
Page 4/15	Item D	Delete '15000' *Insert* '1750'
Page 4/23	Item D	Delete 'Clause G 245' *Insert* 'Clause G 205'
Page 4/25	Item C	In unit column delete 'nr' *insert* 'm'

Please acknowledge receipt by signing the enclosed copy of this letter and immediately returning it to the above address.

Your sincerely

Partner

Signed _____

for _____

Date _____

Figure 6.1 Example of an amendment letter

2.   *Indirect waste* In indirect waste, the material is neither lost
     nor damaged. Rather the material is not used for its intended
     purpose or the quantity used is in excess of what is measured in
     the bill of quantities. Examples of indirect waste are:
     (a) use of facing bricks in lieu of common bricks for substruc-
          ture work; and
     (b) production waste which is the result of work done in excess
          of that specified in the bill of quantities and for which no
          payment is made.

Avoidable waste results in financial loss to the contractor and hence
has an adverse effect on profitability and competitiveness. Therefore,
if a contractor wishes to survive and succeed in business, he or she
should give considerable attention to materials management during
planning stages of the project. In the process, he or she should consider carefully competitive purchasing, the method of delivery, checking material on arrival at site, storage, control, site handling, incorporation into the works, waste minimization, and recovery of unused
materials.

In the age of subcontracting and labour-only subcontracting, it
would be worthwhile for the contractor to discuss his or her measures
on waste with all the subcontractors and ensure the implementation
of these measures during the progress of the works.

Unavoidable waste

This is an incidence of material waste which occurs during the normal course of conversion and cutting to size. The cost of this waste is
borne by the client. However, building works are measured net, that
is, as fixed in position; the wasted material is not measured and the
estimator has to make an allowance for this in his or her rate build-
up.

Waste allowance factor in estimating

There is no standard or hard and fast rules on allowances for waste in
various materials used for building work because there is a general
lack of clear definition of what this constitutes. Therefore, every estimator has to formulate his or her own schedule of allowances depending on his or her skill and experience and hence waste allowance
varies from estimator to estimator and/or from organization to organization. An estimator is able to arrive at the percentage allowance
for material waste through the following

1. *Knowledge of materials wastage* A thorough knowledge and experience of how material waste occurs in construction enables an estimator to make a good assessment and to be able to vary waste allowances in order to reflect:
    (a) type of material;
    (b) position/functional use of material;
    (c) amount of cutting or processing required before fixing of material;
    (d) construction method employed in incorporating material into the building;
    (e) time of year of material usage; and
    (f) general site conditions and their effect on the use of material.

    All these factors have an influence in varying degrees on material waste allowance.

2. *Experience of an organization's materials management policy* If the estimator is aware that the organization in which he or she works has a good materials management policy and that this is successfully implemented, his or her unit rate build-up will reflect this competitive advantage. In such a situation he or she will include very low material wastage allowance in the calculations of the unit rates.

3. *Examination of contract drawings* A careful study of the contract drawings and specification will reveal to the estimator the level of care exercised by the designer to minimize material wastage. For example, a design or specification which does not call for standard sizes of materials indicates a high waste factor as site cutting will be necessary before fixing. As extreme examples, waste allowances can be compared on the following items:
    (a) Prefabricated items (e.g. doors, sanitary appliances, and so on). Here, wastage could be set at a low level, say $2^{1}/_{2}$ per cent. This factor would allow for a one in forty chance of the item being lost through, for example, damage or theft.
    (b) Corrugated roof covering. Wastage here would have to include not only a higher risk of damage, but also for both cutting waste and side and end laps. A waste factor of, say, $33^{1}/_{2}$ per cent would be appropriate in this instance.

PRICING REFERENCES

Several price books (published annually) are available for reference. Some of the more popular books in this range are Spons, Laxtons, Wessex and Hutchins. These books are used by architects, builders and surveyors when either time does not permit a detailed build-up of unit rates or for an indication of prices of work items on which no cost information is available.

Whilst these books are convenient quick price reference points, they only refer to various descriptions of an average job in a specific location and market condition. Hence the user should be careful in their use. To all intents and purposes, the best price reference point for the estimator is solely the one compiled and constantly revised by him- or herself. Armed with the knowledge of the source of the pricing data, the estimator can use this reference point with confidence.

Besides pricing data contained in the price books, they also have the following information:

1.   Professional fees for services provided by architects, quantity surveyors and engineers.
2.   Approximate estimating for various buildings.
3.   Elemental cost planning.
4.   Tables and memoranda.
5.   Brands, trade names and telephone contact numbers.
6.   Basic prices of materials.

ROLE OF COMPUTERS IN ESTIMATING

The estimators of today, like other construction professionals, are obliged to keep pace with the high productivity and quality levels demanded by their organizations in this technological age. As modern projects become larger and more complex, and as construction data increases in quantity but tender periods are not increased to reflect the changes in the construction scene, estimators require technological assistance in order to cope. Hence the present trend is for computers to be employed with increasing effectiveness in the production of estimates.

Whilst it is generally recognized that computers have improved estimating dramatically, it is also accepted that every organization has its own approach to the level of computer involvement in the estimating function. The roles played by computers in the estimating process are numerous but they can generally be said to comprise the following aspects:

1. Expedition of the repetitive aspects of estimators' work and hence better productivity.
2. Efficient recording and storage of data on resources (labour, material and plant) output.
3. Rapid access to and/or retrieval of permanently stored data on resources.
4. Quick processing of unit cost from data stored on file intermixed with revised input cost data.
5. Rapid updating of stored data and subsequent unit cost build-up when required. For instance, logistic constraints may delay the commencement of a project, thereby necessitating the need for revised prices, tender, and so on. If the computer was employed in the estimating process, the production of revised data would present no problem to the estimator.
6. Ease of examining and manipulating the effects of adjustment to cost data.
7. Rapid post-adjudication adjustment to incorporate overhead and profit mark-up.

Generally these and many other computer facilities have improved estimating quality, and at a reduced cost and with minimum risk of errors.

There are also several computer-assisted estimating systems available and each system has diverse modules which can be used; the more commercial systems include a database or library of input prices. However, as bills of quantities have yet to be issued on a magnetic form or disc, the full potential of the computer in estimating has yet to be realized. At the moment the relevant sections of the bill of quantities are transferred into computers by manual entry and are therefore susceptible to human error. To overcome this problem, some computer-assisted estimating systems have software for scanning the printed document and converting the image into an electronic medium for use by computers. This approach is more accurate than the manual copying systems.

The use of computers in estimating, however, should not be seen as a means of replacing the skills of estimators with modern technology. Instead, the computer should be regarded as a 'tool' which assists experienced estimators to make optimum use of their experience and skill in calculating construction costs. This process involves synthesis of unit rates from known or assumed standards of output.

It must also be pointed out that computers are not suitable for estimating alteration and repair works, which require human imagination, judgement, flexibility and adaptability. Moreover, to be beneficial in the estimating process, the cost data stored on computer requires constant updating to reflect and/or incorporate operational feedback from site.

ESTIMATING FOR CIVIL ENGINEERING WORKS

Civil engineering work items are not quantified in any great detail in the bill of quantities. In contrast with that for building works, item descriptions in the bill of quantities for civil engineering works are brief and simple. This brevity and simplicity means that the estimator must look carefully at the drawings and details and study the specification when pricing for civil engineering work.

The estimator may find, for instance, a description for manhole construction as follows:

Manhole; brick (or in situ/precast concrete);
 depth 2–2.5 m .... 4 Nr.

It can clearly be seen that the above description is inadequate for the estimator to price this item without measuring and pricing the various trades in this operation. Moreover the estimator should search for and include all items of operational cost not covered by the short description; for example:

1.   Temporary works which are at the discretion of the estimator.
2.   Design and planning cost for temporary works.
3.   Work items, the measurement of which is at the discretion of the quantity surveyor.

Therefore, estimating for civil engineering works as compared to that of building, is more complicated; it involves more work and requires considerable skill and experience.

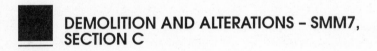

DEMOLITION AND ALTERATIONS – SMM7, SECTION C

ESTIMATING CONSIDERATIONS

Demolition is considered today as work better carried out by specialist contractors who have skilled operatives, suitable equipment and access to markets for salvaged materials. Nevertheless some construction firms have their own demolition departments equipped with all the necessary plant and experienced staff. The factors that influence the pricing of this work include the following.

Type and age of building

The age of a building and type of construction will influence the estimator's judgement in pricing. Taking down an old brick wall built in lime mortar will be easier and cheaper than a new brick wall in cement mortar.

Risks

The higher the risk posed by the demolition to the public and to adjacent properties, the higher will be the demand for safety measures and care required in its execution. Extensive preparatory work and extra precautions slow down the work and hence affect the price.

Working method

This is dictated by the type of work, accessibility, restrictions, and intended use of old materials:

1. *Type of work* It is easier to demolish a whole building than parts of it. If part of an existing building is to be demolished, greater care and safety precautions would be necessary and this would affect the price. Extra expense would be incurred in the provision of tarpaulins, screens and fans to control dust and noise and to protect the public.
2. *Access to main areas* The extent to which mechanical plant (e.g. cranes with crushing weights, winches and compressors) will have access to the main areas of the work for demolition, removal and disposal of the debris.
3. *Restrictions imposed* In the selection of a suitable method of demolition, the estimator will consider all restrictions imposed by the client with regard to noise levels, working hours, and so forth. The magnitude of these restrictions – for example, shorter working hours or noise restrictions leading to selection of a slower labour-intensive demolition method rather than a mechanized method – will affect the price.
4. *Intended use of old material* Where the client intends to reuse the old materials (e.g. facing bricks), the demolition is executed manually with chisel, hammer, and so on, to remove the old

material bit by bit. This is a time-consuming and costly operation.

Property of the old material

If the old materials are to remain the property of the client, the contractor will give no credit for the price of this work. The contractor will only consider giving credit for the old material if it is of value (e.g. lead) and the amount of credit will reflect the salvage value.

Extent of temporary works

Work requiring extensive shoring and other precautionary work will influence the estimator's pricing level. These temporary works can be costly and time consuming to erect, maintain and dismantle.

Disposal and tipping fees

This operation can be expensive where it is necessary to find the right place or method for the disposal of the debris, especially where toxic waste materials are involved.

ALTERATIONS – SPOT ITEMS

Pricing for this type of work requires detailed site measurement, separation of different trades inherent in the measured operation and allowance for temporary works, precautions, and so on, needed to undertake the work. In large projects the labour constants used in preparing the estimates are stated in days rather than hours (e.g. 'X' number of days required for hole-cutting operations followed by 'X' number of days for another operation).

In estimating for alteration works, the estimator adopts a 'spot item' method of estimating. This may be approached in several ways but the most popular methods are as follows:

1. Measuring each trade in an operation and using unit rates in calculating a lump sum estimate for each operation (see method 1 in Example 6.1).
2. Analysing an item description in the bill of quantities into its component operations and assessing the element of cost of each operation in order to arrive at the estimated cost (see method 2 in Example 6.1).

The estimator would include an allowance for the unknown amount of work in each method. For example, provision for making good an unknown quantity of large sections of loose plaster or ceiling that vibrations from operations might shake off.

UNIT RATE BUILD-UP FOR TYPICAL BILL ITEM

Example 6.1

Remove existing softwood double doors, frames and archi-traves from opening size 1,750 x 2,050 mm in 225 mm thick brick walls; block up opening with 65 mm common bricks in cement mortar (1:3); wedge and pin up the top to existing lintel; cut, tooth and bond new brickwork to existing; prepare and plaster both sides of new brick wall; make good to existing plaster; apply one mist coat and two coats of emulsion paint on new plaster; supply and fix softwood skirting to match existing to both sides of new brick wall; make good to existing softwood skirting; knot prime, stop, apply one undercoat and one finishing coat of gloss paint on new softwood skirting; make good all work disturbed.   **Item**

METHOD 1 *Approximate quantities method*

		£
	Remove existing door and frame 1 No @ £7.97 each	7.97
1.75	One brick wall in cement	
2.05	mortar (1:3) 4 m^2 @ £44.50/m^2	178.00
1.75	Wedge and pin up 2 m @ £3.50/m	7.00
2/2.00	Tooth and bond new brick walls to existing 4 m @ £4.50/m	18.00
2/1.75	13 mm plaster to brickwork	
2.05	dubbing average 6 mm, 7 m^2 @ £9.15/m^2	64.05
2/1.75	Prepare and apply two coats	
2.05	of emulsion paint on walls 7 m^2 @ £2.50/m^2	17.50
2/1.75	Softwood skirting including grounds 4 m @ £3.50/m	14.00
2/2	Make good to existing skirting 4 No @ £2.85 each	11.40
2/1.75	Paint on skirting not exceeding 300 mm girth 4 m @ £0.95/m	3.80
	Clean up on completion, labour 1 hour @ £4.72/hour.	4.72

Net Price/Item   **£326.44**

Cont'd

METHOD 2 *Element cost assesment method*
LABOUR

Operation (in sequence of working)	Tradesman's hours	Labourer's hours
(a) Remove door and frame including architrave and remove from site. (1 No)	–	1.50
(b) Build new wall including cutting, bonding, wedging up to existing (4 m²)	11.00	3.50
(c) Dubbing out 13 mm Carlite including making good to existing (7 m²)	7.00	3.50
(d) One mist coat and two coats of emulsion paint on plaster. (7 m²)	1.50	–
(e) Fix softwood skirting and make good existing (4 m)	2.00	–
(f) Prepare and decorate skirting (4 m)	2.00	–
(g) Clean up on completion (Item)	–	1.00
Total Estimated Hours	**23.50**	**9.50**

£

LABOUR

		£
Labourer 9.50 hours @ £4.72/hour	=	44.84
Tradesman 23.50 hours @ £5.53/hour	=	129.96

MATERIALS

		£
(a) Wall, one brick thick, wall 4 m² (say) 500 bricks @ £170/1000	=	85.00
(b) Mortar (allow)	=	6.00
(c) Softwood skirting 4 m @ £3.86/m	=	15.44
(d) Plaster 7 m² (allow) 100 kg Carlite browning	=	18.50
(e) Emulsion paint 7 m² (allow) 2.50 litres @ £15.00	=	15.00
(f) Undercoat 750 ml @ £5.20	=	5.20
(g) Gloss paint 750 ml @ £6.50	=	6.50
Net Price/Item		**£326.44**

Note: In practice there should be a slight difference in price between the two methods used. However, for technical reasons, the calculations have been intentionally manipulated to arrive at the same price for both methods.

GROUNDWORKS – SMM7, SECTION D

Groundworks in a building project comprise excavation, filling and landscaping. The information the estimator requires for pricing the items of work under this section is obtained from the drawings, bill of quantities and specification (where bill of quantities is not used).

The estimator should be aware that the volume of working space required is at the discretion of the contractor. In this respect, he or she should be certain of the adequacy of the amount included in the build up of unit rate to cover the extra cost. He or she should also use his or her skill and experience to determine those items of ground-work that can be effectively excavated by mechanical plant and those by hand.

ESTIMATING CONSIDERATIONS

Factors considered by the estimator when calculating the unit rates for groundworks may include the following.

Location and accessibility of site

The location of site will determine the level of plant usage. A site located in a built-up area with limited accessibility may only be suited to the use of small excavating plants. Where disposal vehicles cannot come alongside excavations for direct loading of excavated material, double handling of excavated material, which slows down the operation, may be encountered.

Ground conditions

The type of soil determines its density, ease of excavation, bulkage and angle of repose. Rock and stiff clay are difficult to excavate and require the use of special expensive plant. Sand or soil that is liable to shear will require expensive earthwork support, that is, close-boarded in lieu of open earthwork support.

Underground services

A site crossed by complex underground services (e.g. water and gas pipes, electric cables, underground drainage) requires extra care in excavation and limits the effective use of plant. This reduces output as the ratio of hand dig increases.

Water table

Ground with a high water table level will hamper excavation and influence the choice of excavating plant. The excavation operation may involve the use of either heavy pumping or sheet piling. Although most of these items may be covered in the bill of quantities, the difficulties of excavating in water will affect the estimator's level of pricing.

Time of year

This refers to the time of year when excavation is programmed to start and the expected weather related to periods. Freezing ground in wintry weather may be difficult to excavate and melting snow and prolonged rainfall requires pumping and removal of slurry from bottom of trenches before concreting. Sizes of timbers for earthwork support will need to be of adequate size and strong enough to withstand the forces acting on them.

Planned ratio of different excavation methods

Proportion of mechanical to hand working envisaged would influence the estimator's approach to pricing. A certain amount of hand digging will be required to bottom up trenches as accurate depths of excavation are difficult to achieve by machine digging only.

Disposal facilities

The size and type of vehicles that are available for the disposal of excavated materials should match the excavating plant. For instance, large excavators should work in conjunction with large-size disposal plant or a fleet of smaller ones to minimize plant waiting time.

Distance to the nearest tip

As surplus excavated material requires to be disposed off site, the distance to tip and tipping charges will influence the level of pricing.

Size of site

The extent of the site will influence the method of excavation adopted. A small restricted site will allow only small excavation plants and limit the number of disposal vehicles. Some restricted sites will not admit excavation plants and hence hand digging may be the only option.

Slope and contours

If the site is steeply sloped, excavation plant and disposal vehicles will be of limited use. Hand digging and transporting of excavated material in wheelbarrows to collection points for disposal vehicles may be a necessity on such projects.

Depth of excavation

Although this will be covered by the Measured Works section of the bill of quantities, problems of excessive depths (e.g. earthwork support, pumping, risks of trench collapse) will influence the estimator's level of pricing.

Proximity to obstructions

The proximity of obstacles on the site (e.g. buildings, substations, trees) will limit the use of plant in the excavation. Plant used on such projects will be difficult to manoeuvre and hence will not be fully utilized, and the rate of hand digging will be high.

Construction programme

The adoption of an expedited construction programme where the contractor is expected to show increased output per period instead of maximization of output per worker per period leads to uneconomic working. This increase in output may be achieved by increasing the labour on site and a prolonged working day. In turn, this could necessitate the use of artificial lighting during shorter winter days and the uneconomic working should influence the estimator's pricing level.

ESTIMATING GUIDELINES

Tables 6.1 and 6.2 give some estimating guidelines for hand excavating and machine excavating respectively. In Table 6.2 it should be noted that the outputs indicated per machine size are strictly only a guide since machine output can be influenced by a variety of factors, as follows:

1. Skill of the machine operator.
2. Effectiveness of the machine selected for the work.
3. Whether the machine works continuously or intermittently.
4. Nature of soil to be excavated and depth of excavation.

5. Existence of obstructions above and below ground, and underground services requiring care and attention.
6. Distance travelled by machine to dispose of excavated material.
7. Weather conditions (e.g. too cold, hot, wet or windy).

Table 6.1 Guide to average labour output for hand excavating and filling

Hand excavating description	Unit	Labour hours
Excavating to reduce levels		
Maximum depth not exceeding 1.00 m	m³	2.40
Excavating basement		
Maximum depth not exceeding 1.00 m	m³	2.40
Maximum depth not exceeding 2.00 m	m³	2.90
Excavating trench width not exceeding 0.30 m		
Maximum depth not exceeding 1.00 m	m³	2.75
Maximum depth not exceeding 2.00 m	m³	3.50
Fixing earthwork support	m³	18.00
Striking earthwork support	m³	9.00
Disposal of excavated material		
On-site not exceeding 100 m distance	m³	1.00
Filling to excavation average thickness		
over 0.25 m, material obtained off-site	m³	1.63

Note: Output based on firm soil in fair working weather.

Table 6.2 Mechanical excavating – guide to machine output

Type of excavation	Output of machine (m³/hr)			
	0.25 m³ size bucket	0.38 m³ size bucket	0.50 m³ size bucket	0.75 m³ size bucket
Reduce levels	6	7	9	14
Trenches	7	8	11	17
Pit	8	11	–	–
Basement	8	9	12	18

CHOICE OF MECHANICAL EXCAVATOR

The estimator must consider the following points carefully before embarking on pricing excavation by machine.

Plant to be used

Plant used in excavation operations vary in kind and size and the selection of the correct type and size of plant depends largely on the sort of excavation involved. Table 6.3 shows types and uses of excavating plant.

Table 6.3 Mechanical excavators	
Plant	*Type of operation*
Face Shovel	Bulk excavation, generally cutting banks
	Basement excavation
	Loading excavated material
	General excavation above track level
Skimmer	Excavations to reduce level
	Roadworks
Dragline	Excavation to reduce level
	Deep excavation to basements
	Cleaning streams, ditches, and so on
	General excavation below track level
Backactor	Drain and surface trench excavation
	Basement excavation
Crawler tractor shovel	General site excavation
	Loading of disposal vehicles
Tractor scrapers	Bulk surface excavation and removal
Tractor-mounted shovels	Over site and surface excavation
	Trench excavation
Bulldozer	Reducing levels over site
	Filling and grading
	Forming embankments

Bulkage of the soil

The estimator must be able to assess the anticipated bulking of excavated material. The extent of bulking will determine the amount of excavated material that can be disposed of by spoil disposal vehicles.

Table 6.4	Soil bulkage factors

Type of Soil	Increase in bulk (%)
Loose	$12^1/_2$
Firm	25
Compact	25
Clay	$33^1/_3$
Chalk	$33^1/_3$
Rock	50

Table 6.4 is a guide to the bulkage to be expected from various types of soil.

EARTHWORK SUPPORT

This refers to construction equipment required temporarily or permanently to uphold the sides of excavation during the execution of groundworks. Generally this operation is regarded as the contractor's risk. However, he or she is given the opportunity to decide whether or not to price for this item in the bill of quantities. It is therefore one of the estimator's tasks to decide if it will be required, and for what length of time.

The following factors will influence the estimator's method of pricing earthwork support.

Stability of ground

The nature of soil will determine the type of earthwork support required. Excavation in loose and sandy soil requires close-boarded earthwork support while open earthwork support will be required for excavation in firm soil.

Depth and width of excavation

While some shallow excavations may require no earthwork support, to facilitate safe working there is a need for support in all deep excavations. The depth of excavation determines the size of timber members, and in some cases the use of steel sheets may be necessary. The wider the excavation the larger the section of struts or the use of steel sections.

Thrust of adjoining buildings

Pressures exerted by adjoining buildings on excavations vary and at times can be troublesome. It normally requires the employment of larger sections of timber and/or steel sections to overcome the problems that these can cause.

Possibility of disturbance from vibration

Vibration from adjoining roads can exert a lot of pressure on the earthwork support and its effective functioning. This may yet again require larger sections of timber and/or steel sections.

The number of uses envisaged

The greater the number of uses that can be obtained from the timber sections, the more economic they become. Therefore, the contractor should, wherever possible, ensure that timber sections receive maximum number of use.

Period excavation is likely to remain open and supported

When the excavation is expected to remain open for long periods of time, the earthwork support will require larger timber sections or steel sections to ensure continued stability. The number of uses of the earthwork support in this instance will diminish and therefore should be reflected in the estimator's method of pricing.

UNIT RATE BUILD-UP FOR TYPICAL BILL ITEMS:

Example 6.2

Excavating to reduce levels; maximum depth not exceeding 1.00 m.	m^3

DATA
(i)   Firm soil to be excavated
(ii) Mechanical excavation
(iii) Machine output – 14 m^3/hour
(iv) Basic rates:
         Machine and operator   £18.50/hour

SYNTHESIS	£	£
Machine and operator @ £18.50/hour	18.50	
Standing time (allow) 25%	4.63	
Cost/hour	23.13	

Machine output = 14 m^3/hour

Therefore cost/1 m^3 = $\dfrac{£23.13}{14}$ 1.65

Net unit rate per m^3   **£1.65**

Example 6.3

Excavating basement and the like; maximum depth not exceeding 2.00 m.   **m³**

DATA
(i)   Firm soil to be excavated
(ii)   Mechanical excavation with a banksman in attendance
(iii) Machine loading direct to disposal vehicle
(iv)   Machine output – 18 m³/hour
(v)   Machine bucket efficiency – 67%
(vi)   Basic rates:
         Banksman   £4.72/hour
         Machine and operator   £18.50/hour

SYNTHESIS	£	£
Machine and operator @ £18.50/hour	18.50	
Banksman @ £4.72/hour	4.72	
	23.22	
Standing time (allow) 25%	5.81	
Cost/hour	29.03	

Machine output = 18 m³/hour
  x 67% bucket efficiency = 12.06 m³

Therefore cost/1 m³ = $\dfrac{£29.03}{12.06}$ 2.41

Net unit rate per m³   **£2.41**

Example 6.4

Excavating trenches; width over 0.30 m; maximum depth not exceeding 2.00 m. m³

DATA
(i)   Firm soil to be excavated
(ii) 80% mechanical excavation and 20% hand excavation
(iii) Machine loading direct to disposal vehicle
(iv) Machine output – 8 m³/hour
(v)   Labour output – hand excavation, 1 m³/3.50 hours
(vi) Basic rates:
      Labourer £4.72/hour
      Machine and operator £18.50/hour

SYNTHESIS		£	£
(i)	Proportion of machine excavation (80%)		
	Machine and operator @ £18.50/hour	18.50	
	Standing time (allow) 25%	4.13	
		————	
	Cost/hour	22.63	
		═══	

Machine output = 8 m³/hour

Therefore cost/1 m³ = $\dfrac{£22.63}{8}$ = £2.83/m³

	80% machine excavation = £2.83 x 0.8	2.26	
(ii)	Proportion of hand excavation (20%)		
	Labour output = 1 m³/3.50 hours		
	Cost/1 m³ = 3.50 hours x £4.72/hour = £16.52		
	20% hand excavation = £16.52 x 0.2	3.30	
		————	
		5.56	
		═══	

Net unit rate per m³ **£5.56**

Example 6.5

Earthwork support; distance between opposing faces not exceeding 2.00 m; maximum depth not exceeding 2.00 m.   m²

DATA
(i)   Firm soil to be supported
(ii)   Material in timber members and metal struts
(iii) Utilization factor of 12
(iv)   Excavation to remain supported for two weeks
(v)   Labour output:

Unloading timber	2 hours/m³
Fixing timber	18 hours/m³
Stripping timber	9 hours/m³

(vi)   Basic rates:

Timber	£155.00/m³ delivered
Timberman	£ 5.04/hour
Carpenter	£ 5.53/hour
Labourer	£4.72/hour

(vii) Section of trench taken 10 metres long
       x 1 metre deep using one 150 x 75 mm
       timber waler/side; 150 x 50 mm poling
       boards at 1 metre centres and metal trench
       struts at 2 metres centres

SYNTHESIS   £   £
(i)   Materials

Waler 2 sides x 10 m x 150 x 75 mm	=	0.225 m³
Poling boards 2 sides x 11 number		
x 150 x 50 mm	=	0.165 m³
		0.390 m³

0.39 m³ timber @ £155/m³ delivered	60.45
Unload 0.39 m³ @ 2 hour/m³ = 0.39 x £9.44	3.68
	64.13
Allow for waste (7^1/$_2$%)	4.81
	68.94

Allow 12 number of uses

Therefore 1 number use = $\dfrac{£68.94}{12}$

                                                    C/F   5.75

	B/F	5.75
Nails and fixings per use (allow)		0.50
		6.25

Hire of trench struts for two weeks @
£1.25/week each 6 number x £1.25 x 2 weeks   15.00

                                                21.25

(ii) Labour
Fix timber 18 hours/m^3
Strip timber 9 hours/m^3
= 27 hours x 0.39 m^3 x £5.53   = £58.23

Fix and strip trench struts
6 No at 0.10 hour each @ £5.53/hour =   £3.32

                                                  61.55

(iii) Extra excavation and backfilling for earthwork
support

Waler   75 mm
Poling   50 mm
           125 mm x 2 sides = 250 mm thick

Trench
10 m long x 1 m deep x 0.25 m thick = 2.5 m^3

Excavate 2.5 m^3 @ £ 5.36/m^3 = £13.40
Fill 2.5 m^3 @ £13.94/m^3 = £34.85

                                                48.25

                                                131.05

Total area 10 m x 1 m x 2 sides = 20 m^2
20 m^2 cost £131.05

Therefore cost of 1 m^2 = $\dfrac{£131.05}{20}$ 6.55

Net unit rate per m^2   **£6.55**

Example 6.6

Earthworks support; distance between opposing faces not exceeding 2.00 m; maximum depth not exceeding 4.00 m.   m²

DATA
(i)   Sandy soil to be supported
(ii)   Material in timber members and metal struts
(iii) Utilization factor of 8
(iv)   Excavation to remain supported for two weeks
(v)   Labour output (as Example 6.5)
(vi)   Basic rates (as Example 6.5)
(vii) Section of trench taken 10 metres long x 3 metres deep
         using 30 mm timber boards; three 150 x 75 mm timber
         walers/side; 150 x 50 mm poling boards at 1 metre
         centres and metal trench struts at 2 metre centres

	SYNTHESIS			£	£
(i)	Materials				
	Waler 2 sides x 10 m x 150 x 75 m x 3	=	0.675		
	Poling boards 2 sides x 11 number x				
	3 m deep x 150 x 50 mm	=	0.495		
	Timber boards 2 sides x 10 m x				
	3 m x 30 mm	=	1.800		
			2.970 m³		

2.97 m³ timber @ £155/m³ delivered		460.35
Unload 2.97 m³ timber @ 2 hours/m³ = 2.97 x £9.44		28.04
		488.39
Allow for waste (7^1/2%)		36.63
		525.02

Allow 8 number of uses

$$\text{Therefore 1 number use} = \frac{£525.02}{8} \qquad 65.63$$

Nails and fixings/use (allow)	0.50
C/F	66.13

                                              B/F   66.13

Hire of trench struts for two weeks @
    £1.25/week each = 18 number x £1.25
    x 2 weeks   45.00
                                                   ───────
                                                   111.13

(ii)   Labour
        Fix timber 18 hours/m^3
        Strip timber 9 hours/m^3
            = 27 hours x 2.97 m^3 x £5.53   = £443.45

        Fix and strip trench struts
        8 No at 0.10 hour each @ £5.53/hour =   £9.95
                                                 ───────
                                                          453.40

(iii)   Extra excavation and backfilling for earthwork
            support

        Waler   75 mm
        Poling   50 mm
        Boards   30 mm
                   ──────
                   155 mm x 2 sides = 310 mm thick

        Trench
        10 m long x 3 m deep x 0.31 m thick = 9.3 m^3

        Excavate 9.3 m^3 @ £5.36/m^3 =   £49.85
        Fill 9.3 m^3 @ £13.94/m^3 = £129.64
                                                  ────────
                                                          179.49
                                                          ───────
                                                          744.02
                                                          ═══════

        Total area 10 m x 3 m x 2 sides = 60 m^2
        60 m^2 cost £744.02
        Therefore cost of 1 m^2 = $\dfrac{£744.02}{60}$ 12.40
                                                          ───────

Net unit rate per m^2   **£12.40**
                                                          ═══════

Example 6.7

Disposal of excavated material; off site.

m³

DATA
(i)   Excavated material loaded direct to
         10 m³ capacity disposal vehicle
(ii) Cost of excavating machine and loading
         included in excavating unit rates
(iii) Bulking of excavated material 25%
(iv) Speed of disposal vehicle 30 kilometres/hour
(v) Tip 20 kilometres away from site
(vi) Basic rates:
         Disposal vehicle and driver   £15/hour
         Tipping fees   £8.50/load

SYNTHESIS   £   £

Speed of disposal vehicle 30 kilometres/hour
Therefore 1 kilometre in 2 minutes

Total round trip time
20 kilometre travel = 20 x 2   =   40 minutes
Tipping and standing time   =   15 minutes
Loading time   =   60 minutes
                                                  115 minutes
                                                  1.92 hours

Capacity of disposal vehicle   =   10 m³

Actual load/trip   = $\dfrac{10 \text{ m}^3 \times 100\%}{125\%} = 8 \text{ m}^3$

Cost of disposing 1 m³   = $\dfrac{£15.00 \times 1.92 \text{ hours}}{8 \text{ m}^3}$ 3.60

Therefore cost of tipping 1 m³   = $\dfrac{£8.50}{8 \text{ m}^3}$
                                                                                           1.06
                                                                                     _____
                                                                                           4.66
                                                                                     ========

Net unit rate per m³   **£4.66**

Example 6.8

Filling to make up levels; material obtained off site; quarry rejects; average thickness over 0.25 m. m³

DATA
(i) Hardcore tipping directly to required position
(ii) Hardcore spread and rolled by hand
(iii) Labour output – spreading and levelling 2.2 hours/m³
(iv) Allowance of 20% required to compensate
     for voids after consolidation
(v) Basic rates:
     Hardcore £ 9.50/m³ delivered
     Rammer hire rate £1.25/hour
     Rammer operator £4.86/hour

SYNTHESIS	£	£
(i) Materials		
1 m³ hardcore @ £9.50/m³ delivered	9.50	
Allow 20% compaction	1.90	
	11.40	
Allow for waste (2½%)	0.29	
	11.69	
Cost/m³ hardcore	11.69	
(ii) Labour and plant		
Spreading and rolling/m³		
Labour 0.40 hour @ £4.86/hour = 1.94		
Rammer 0.25 hour @ £1.25/hour = 0.31		
	2.25	
	13.94	

Net unit rate per m³ **£13.94**

CONCRETE WORK – SMM7, SECTION E

Concrete work consists of mixing specified proportions of cement, water and aggregates, transporting the mix and placing it in a required location. When cured at a recommended temperature and humidity, concrete solidifies and attains the required strength.

ESTIMATING CONSIDERATIONS

Factors to be considered by the estimator when building up unit rates for concrete works may be summarized as follows.

Type of concrete

The estimator will examine the economies of the use of ready-mixed and site-mixed concrete. Mixing concrete on site requires space for installation of mixing plant. For congested sites it is advantageous to employ ready-mixed concrete. However, ready-mixed concrete is delivered in minimum loads by the loose cubic metre; therefore an allowance must be made for consolidation, shrinkage, waste and spillage. Where small volumes of concrete are required and the site can accommodate a mixing plant, then site mixing is the most economic option.

Location of site mixing plant

When site-mixed concrete is used, positioning mixing plants close to the works enables concrete to be transported in wheelbarrows. This arrangement will cut down transporting time and cost, and will lead to a competitive pricing level. Hence the estimator should study the site layout carefully before the unit rates build up.

Method of site mixing

Concrete may be either hand- or machine-mixed. It is, however, more economical to use machine-mixed concrete on projects where a large volume of concrete is required. Machine mixing is carried out by one or more of the different types and sizes of mechanical mixers available. Mixers can be hand or mechanically loaded. Moreover there are others which are loaded by means of overhead hoppers and a gravity feed. The size to be selected will depend on the output required at a given time.

Method of transporting concrete

The cheapest method of transporting concrete over the site will be considered by the estimator. Wheelbarrows are employed on small projects where distance between concrete mixing and placing points can be short. However, on large projects, concrete can be transported by pumping but where there is a long distance to be covered, dumpers and lorries are employed.

Method of hoisting

In multistorey structures projects, the cost of tower cranes and hoists for lifting the concrete is generally included in the preliminaries section of the bill of quantities. However, the location of the concrete-mixing plant and the lifting plant in relation to the works should be planned so as to enhance the effectiveness of their utilization. This should aim at the best allocation of lifting plant time to all items of work on site.

Placing and consolidation

When concrete reaches its final placing position, the transporting vehicle assisted by concrete placers discharges the concrete to its final position. The placers take delivery and then work and consolidate the concrete around reinforcing bars using a poker vibrator. The attendance of concrete placers is still required where concrete is transported by pump, as in the use of ready-mixed concrete.

Curing the concrete

All vertical and horizontal surfaces of placed concrete require curing. The curing process requires keeping the wet concrete continuously damp for some days. This can be achieved by covering the wet concrete with sand, polythene sheeting or hessian fabric.

Location of pour

Concreting in confined space creates difficulties in placing and curing the concrete, hence the estimator should make an allowance for this construction problem in his or her pricing.

Time of year for concreting

Concrete work carried out in winter months requires additional precautions to prevent concrete from freezing before setting. There may be a need to warm the aggregates, water, and so on, used in mixing the concrete. During hot summer months, concrete needs protection from direct sunlight and, sometimes retarding agents added to the mix to prevent flash setting.

Choice of cement

Cement may be purchased either in 50 kg bags or in bulk. The cement purchased in bags must be carefully stored in watertight site huts and those purchased in bulk stored in cement silos located next to the concrete mixer. Cement purchased in bags is ideal for small projects but costs more as a result of unloading and site storage.

Shrinkage factor of aggregates

The percentage of void created in concrete depends on the type and sizes of coarse aggregates used in the mix. Therefore the estimator must be conversant with the percentage of voids created by the use of various types and sizes of coarse aggregates and allow for them in his or her pricing. Table 6.5 is a guide on percentage addition for voids in various coarse aggregates:

Table 6.5   Aggregate shrinkage factors

Type of aggregate	Addition (%)
Gravel 10–25 mm	50
Gravel 38–75 mm	45
Crushed stone	50
All-in ballast	40

Concrete mixing plants

There are various types of mixers on the market and the estimator should select one on which to base his or her pricing. Table 6.6 gives an example of the cost of mixing concrete using $5/3\frac{1}{2}$, 7/5, 10/7 and 14/10 size concrete mixers. Concrete mixer size $5/3\frac{1}{2}$ denotes a mixer in which when loaded with 5 m^3 of dry concrete, the concrete shrinks to $3\frac{1}{2}$ m^3 after mixing. By the same token, 7 m^3, 10 m^3 and 14 m^3 of dry concrete in concrete mixer size 7/5, 10/7 and 14/10 shrinks to 5 m^3, 7 m^3 and 10 m^3 respectively after mixing.

Table 6.6   Cost of mixing concrete by machine

Description	Size of Mixer*			
	5/3½	7/5	10/7	14/10
Assumed output/hour in m³	1	2	3	4
Attendance				
Driver	1	1	1	1
Labourers	2	2	3	2
Fixed cost/work				
Hire charge	25.40	28.75	32.15	37.27
Running cost/work				
Driver 40 hours @ £4.86	194.40	194.40	194.40	194.40
Cleaning, etc. 5 x 2 x ½ x £4.86	24.30	24.30	24.30	24.30
Fuel				
40 x 1½ x 42p	25.20	–	–	–
40 x 2 x 42p	–	33.60	–	–
40 x 2½ x 42p	–	–	42.00	–
40 x 3 x 42p	–	–	–	50.40
Lubricating oil and grease, say	2.00	2.25	2.50	2.75
Weekly cost of running machine	271.30	283.30	295.35	309.12
Hourly cost of machine assuming 75% Capacity; it produces concrete for 30				
hours	9.04	9.44	9.85	10.30
Labourer				
2 @ £4.72	9.44	9.44	–	9.44
3 @ £4.72	–	–	14.16	–
Hourly cost of mixing concrete	18.48	18.88	24.01	19.74
Cost of mixing 1 m³	18.48	9.44	8.00	4.94

*The sizes were originally based on actual cubic feet volumes.

Average output

Table 6.7 gives a guide to average output for concrete work.

Table 6.7 Guide to average labour output for concrete work		
Description	*Unit*	*Labour hours*
Unloading and storing bagged cement	tonne	1.00
Mixing concrete by hand	m³	5.53
Wheeling, depositing and returning average 25 m	m³	1.33
Taking delivery of ready-mixed concrete	m³	0.33
Placing and compacting concrete		
Foundations	m³	2.33
Ground beams	m³	6.00
Slab thicknesses not exceeding 150 mm	m³	4.00
Slab thicknesses 150–450 mm	m³	6.00
Wall thicknesses not exceeding 150 mm	m³	4.00
Wall thicknesses 150–450 mm	m³	4.00
Beams	m³	7.00
Columns	m³	9.00

FORMWORK

Formwork cost is related to a number of uses that can be obtained from the timber. The utilization factor is dependent on the number of repeats possible from the architect's design and contract programme. Although factors such as cleaning, repairing, use of shutter-release agents (which prolongs the life of formwork) can be time-consuming and expensive, it is still economical if the maximum number of uses can be obtained. Therefore, before pricing the works, the estimator should study various concrete sections shown on tender drawings to ascertain the number of uses that can be obtained from formwork.

The factors that may influence the estimator's pricing methods may include the following.

The weight of concrete when poured

The pouring of large volumes of concrete subjects formwork to great pressure and therefore calls for additional and stronger supporting members (e.g. beam hangers, wall ties, column cramps, adjustable props and stronger forms).

Type of formwork material
There is metal, timber and proprietary formwork on the market and the estimator should make the most economic choice.

Whether formwork is subject to reuse
The estimator's price should reflect the reduced utilization factors for permanent formwork which does not allow for repeated usage.

Fixing and stripping per use
The output for fixing and stripping formwork depends on the experience and motivation of contractor's operatives. The estimator should therefore rely on his or her company's historical labour output for this information.

Numbers of uses envisaged
The formwork price will be reflected by the utilization factor after taking into account the labour cost of cleaning, repairs and oil treatments. As explained above, the higher the utilization factor the more competitive is the formwork rate.

Time of year for concreting
The concreting period will affect the fixing and striking times as well as the number of uses that can be obtained from the formwork. Hardening time for concrete in winter is long and hence prolongs formwork striking time of concrete placed during the winter months.

Complexity of design
Complex design requires complex formwork which consumes timber in fabrication and reduces the repeated usage of formwork. The estimator should consider this factor in his or her pricing.

Designed concrete finish

Where the concrete is designed to achieve an intricate finish this will affect the design and fabrication of the formwork as well as its utilization factor.

Average output

Table 6.8 gives a guide to average labour output for formwork.

Table 6.8 Guide to average labour output for formwork		
Description	Unit	Labour hours
Fabricate plywood formwork	m^2	0.80
Fabricate softwood framing	m^3	27.36
Fixing and striking walls	m^2	1.50
Fixing and striking soffits	m^2	1.50
Fixing and striking columns and beams	m^2	2.50
Cleaning and applying mould oil	m^2	0.30

REINFORCEMENT

The current practice in this operation is that suppliers undertake the fabrication of reinforcement bars to a structural engineer's drawings. Hence the reinforcing bars arrive on site already cut to lengths, bent, bundled and clearly marked to facilitate sorting out and fixing operations. Notwithstanding the above, the estimator should make an allowance in his or her rate to cover tying wire, spacers and rolling margins. Table 6.9 gives a guide to average labour output for bar reinforcement.

Table 6.9 Guide to average output for bar reinforcement

Description	Steel Fixer Hours (per tonne)						
	Size of steel reinforcement (mm)						
Sorting, stacking, cutting, bending and fixing as required	6	8	10	12	16	20	25
Foundations	90	84	70	60	55	50	45
Beds	100	84	70	60	55	50	45
Floors and roofs	105	85	72	62	58	53	50
Walls	120	105	85	75	70	60	55
Beams and columns	125	106	90	80	70	65	55

Unloading, sorting, stacking, cutting, fixing, etc., straight bars	75% of hours shown above
Unloading, sorting, stacking, fixing etc., ready-cut and bent bars	50% of hours shown above
Unloading steel reinforcement	2 hours per tonne

UNIT RATE BUILD-UP FOR TYPICAL BILL ITEMS:

Example 6.9

Plain *in situ* concrete (1:2:4–20 mm aggregates); foundations; poured on or against earth.	m^3

DATA
(i)   Site mixed using 10/7 mixer
(ii)   Concrete wheeled by barrows to placing position 25 m away
(iii) Basic rates:
       Cement (bagged) £56.00/tonne delivered
       Sharp washed sand £7.00/m^3 delivered
       20 mm coarse aggregates £7.00/m^3 delivered
       Labourer £4.72/hour

SYNTHESIS

	£	£
1 tonne bagged cement @ £56/tonne delivered	56.00	
Unload 1 hour @ £4.72/hour	4.72	
	60.72	
1 m^3 cement @ £60.72/tonne x 1.442 tonne/m^3	87.56	
2 m^3 sand @ £7.00/m^3	14.00	
4 m^3 aggregates @ £7.00/m^3	28.00	
	129.56	
Allow for voids (40%)	51.82	
Cost of 7 m^3 concrete	181.38	

Therefore cost of 1 m^3 concrete = $\dfrac{£181.38}{7}$ **25.91**

		£
Cost of mixing 1 m^3 concrete (from Table 6.6)		8.00
Wheeling 1 m^3 concrete	2.33 hours	
Placing 1 m^3 concrete	1.33 hours	
	3.66 hours @ £4.72/hour	17.28
		51.19
Allow for waste (2^1/$_2$%)		1.28
Net unit rate per m³		**£52.47**

Example 6.10

Reinforced in-situ concrete (1:2:4–20 mm aggregate); columns.
$$m^3$$

DATA
(i)   Site mixed using 10/7 mixer
(ii) Concrete wheeled by barrows to placing position 25 mm away
      and placed by hoist
(iii) Basic rates (as Example 6.9)

SYNTHESIS		£	£
Cost of 1 m³ cement (as Example 6.9)		25.91	
Cost of mixing 1 m³ concrete (from Table 6.6)		8.00	
Wheeling 1 m³ concrete	1.33 hours		
Placing 1 m³ concrete	4.00 hours		
	5.33 hours @ £4.72/hour	25.16	
		59.07	
Allow for waste (2¹/₂%)		1.48	
		60.55	

Net unit rate per m³   **£60.55**

Example 6.11

Formwork; soffit of slabs; horizontal; slab thickness not exceeding 200 mm, height to soffit 1.50–3.00 m.	m²

DATA
(i) Formwork design 19 mm plywood supported
   on acrow props including horizontal members
(ii) Timber bearers used at an average rate of
   0.05 m³ to every metre square of plywood
(iii) 1 Number acrow prop/square metre of
   formwork and in use for four weeks
(iv) Utilization factor of 6
(v) Basic rates:
   Timber £155.00/m³ delivered
   19 mm plywood   £6.35/m² delivered
   Acrow props   £1.25 each/week
   Carpenter   £5.53/hour
   Labourer   £4.72/hour

SYNTHESIS

	£	£
(i) Materials		
Timber £155.00/m³ delivered	155.00	
Unload 2 hours/m³ @ £4.72/hour	9.44	
	164.44	
0.05 m³ timber @ £164.44/m³	8.22	
1 m² 19 mm plywood £6.25/m² delivered	6.25	
Unload (allow)	0.10	
	14.57	
Allow for waste (7½%)	1.09	
	15.66	
Labour fabrication 0.80 hour/m² @ £5.53/hour	4.42	
6 number of uses	20.08	

Cont'd

Therefore cost of 1 use = $\dfrac{£20.08}{6}$ 3.35

(ii)   Labour
       Fix and strip   1.50 hours/m²
       Clean, repair,
           apply mould oil   0.30 hour/m²
                              1.80 hours @ £5.53/hour   9.95

       1 number acrow prop @ £1.25 x 4 weeks   5.00

       Bolts, nails and sundry material (allow)   0.25

       Mould oil (allow)   0.25

       Repair materials (allow)   0.25   15.70

Net unit rate per m²   **£19.05**

Example 6.12

Formwork; column; isolated, rectangular shaped; regular   m²

DATA
(i)   Column size 300 x 250 mm and 3.50 m high
(ii) Design of formwork to be 19 mm thick plywood and 50 x 50 mm
        softwood framing 9 number column clamps and 4 number
        steel props
(iii) Utilization factor of 8
(iv) The length of time allowed for the formwork – 2 weeks/column
(v)   Basic rates:
        Timber and plywood (as Example 6.11)
        Column cramps £0.75 each/week
        Steel props £1.25 each/week
        Labour rates (as Example 6.11)

SYNTHESIS   £   £
(i)   Materials
      19 mm plywood   2 x 0.319 x 3.50 = 2.23
                       2 x 0.269 x 3.50 = 1.88
                                            _____

                                            4.11 m²
                                            ========

      50 x 50 mm softwood framing
      4 x   2 x 3.50 m = 28.00
      2 x 10 x 0.238   =   4.76
      2 x 10 x 0.188   =   3.76

                    ~~36.52~~ m x 0.05 x 0.05
                         = 0.09 m³

      4.11 m² plywood @ £6.25/m² delivered   25.69
      Unload (allow)   0.40
      0.09 m³ timber @ £155/m³ delivered   13.95
      Unload 0.09 m³ timber @ 2 hours/m³ = 0.09 x £9.44   0.85
                                                        _____

                                                         40.89

      Allow for waste (7^1/2%)   3.07
                                                        _____

      Cost of 4 m² formwork   43.96
                                                        ========

Cont'd

Therefore cost of 1 m^2 = $\dfrac{£43.96}{4}$ 10.99

(ii)   Labour
Fabricating plywood 0.8 hours @ £5.53/hour   4.42
Fabricate softwood framing 0.09 m^3 for area of 4 m^2

Therefore 0.0225 m^3 @ 27 hours/m^3 @ £5.53/hour   3.36

Cost of 8 number of uses   18.77

Therefore cost of 1 use = $\dfrac{£18.77}{8}$ 2.35

(iii)   Labour
Fix and strip   2.5 hours/m^2
Clean, repair, apply mould oil   0.3 hours/m^2
                                   2.8 hours
                                      @ £5.53/hour   15.48

9 No clamps @ £0.75/week x 2 weeks   13.50
4 No steel props @ £1.05/week x 2 weeks   8.40

Cost of 4 m^2 of formwork   21.90

Therefore cost of 1 m^2 formwork = $\dfrac{£21.90}{4}$ 5.48

Bolts, nails, sundry material (allow)   0.50

Mould oil (allow)   0.25

Repair material (allow)   0.25   6.48

Net unit rate per m^2   **£24.31**

Example 6.13

**Bar reinforcement; 12 mm diameter; straight; horizontal;
length 12.0–15.00 m; in beams.**   **tonne**

DATA
(i) Bar reinforcement delivered to site cut to
     various sizes, bent and labelled
(ii) Waste and rolling margin $7^1/_2$%
(iii) Basic rates:
       12 mm diameter mild steel bar reinforcement
          £525.00/tonne delivered
       Tying wire   £4.50/kg delivered
       Plastic spacers   £0.10 each
       Steel fixers   £5.53/hour

SYNTHESIS

	£	£
1 Tonne 12 mm diameter bars @ £525/tonne delivered	525.00	
Unload 2 hours/tonne x £4.72/hour	9.44	
	534.44	
Allow for waste ($7^1/_2$%)	40.08	574.52
Steel fixers 80 hours @ £5.53/hour	442.40	
Tying wire 15 kg @ £4.50/kg	67.50	
Plastic spacers 200 number @ £0.10 each	20.00	529.90
Net rate per tonne		**£1 104.42**

MASONRY – SMM7, SECTION F

Work items considered under this section are brickwork, blockwork and damp-proof course.

COST OF BRICKWORK

The components of cost of one metre square of brickwork are the cost of bricks, mortar, and bricklayers' and labourers' wages. However, this cost may be influenced by the following factors.

Size of bricks

Although bricks vary in size, the most common brick size is 215 x 102.5 x 65 mm. As bricklayers are familiar with this brick size, the estimator should include learning-curve cost in his or her pricing where bricks specified are of different sizes from the most common bricks. Other available brick sizes are: 50 x 102.5 x 215 mm; 75 x 102.5 x 215 mm and 90 x 100 x 190 mm.

Weight of bricks

On average, a standard brick weighs 3.95 kg and has compressive strength of 34.5 N/mm^2. Bricks that weigh more than this subject bricklayers to extra strain and affects their performance.

Bond

Bricks are laid in one of the following common bonds: header, stretcher, English garden and Flemish garden wall. The number of bricks per square metre in each of the above bonds vary and so does the bricklayer's output.

Amount of mortar used

The amount of mortar used depends on whether the bricks have frogs and the size of frogs. Frogs affect the amount of mortar required and should influence the estimator's pricing.

Thickness of work

Material content in one-brick wall is more than that in a half brick wall (see Table 6.10) and so is the labour input. Hence one brick wall should cost more than a half-brick wall.

Table 6.10 Number of bricks/m²				
Size of bricks	Thickness of joints	Wall thickness		
		¹/₂ brick 102.5 mm	1 brick 215 mm	1¹/₂ brick 327.5 mm
65 mm	10 mm	59	118	117

Type of mortar

Gauged mortar is easier to work with than cement mortar and hence the output of bricklayers working with gauged mortar is more than those using cement mortar under the same working conditions.

Quality of workmanship

A higher quality of workmanship is expected in laying facing bricks than that in laying common bricks which are to be subsequently covered. More time is required in high-quality work in bricks selection, and extra care in handling, laying, pointing and protecting the finished work. All these are expensive operations which the estimator should take into account.

Location of work

Straightforward work in a good accessible location should have a standard output. The operator's output is affected where work is complex or located in an awkward position.

Cost of bricks

Bricks for large contracts are purchased in thousands and delivered to site in plastic wrapping, bands and pallets. However, small quantities can be purchased for small contracts. The cost of unloading and stacking the bricks is added to the cost of brickwork. Bricks may be unloaded by various methods, each with a different cost:

1. *By hand* For engineering, facing and special bricks which arrive on site loose and require care in handling to avoid damage.
2. *By crane or other mechanical methods* These mechanized unloading methods are ideal for bricks which arrive on site with loose packaged or palletized bricks.

The time taken to unload and stack bricks is influenced by:

1. *Method used* Faster if mechanized methods adopted.
2. *Degree of care necessary* As in the case of engineering and facing bricks.
3. *Position of stack in relation to delivery vehicle* The longer the distance the longer the time taken to unload.
4. *Site conditions* Untidy or congested site and windy or wet conditions hamper the unloading operation.
5. *Size and weight of bricks* The larger the size and the heavier the weight the smaller the number that can be unloaded at any moment in time.

Cost of mortar

Bricks are bonded together by various types and mixes of mortar according to specification. The usual mortar mixes are cement mortar 1:3 or 1:4 and cement-lime mortar 1:1:6 or 1:2:9. Brick mortars are normally mixed on site; however, ready-mixed mortar can be purchased if required. A 65 mm-deep brick is laid in 10 mm thickness of mortar joints. (See Tables 6.11 and 6.12 for costing guides.)

Table 6.11 Quantity of mortar/m² of brickwork

Thickness of joints	Size of bricks	Frogs	Thickness of wall		
			¹/₂ brick 102.5 mm	1 brick 215 mm	1¹/₂ brick 327.5 mm
10 mm	65	1	0.03 m³	0.07 m³	0.11 m³
10 mm	55	1	0.035 m³	0.08 m³	0.125 m³
10 mm	75	1	0.028 m³	0.066 m³	0.104 m³

Note: The quantity of mortar/vertical joint of each brick size is 0.01 m³.

Table 6.12   Cost of mortar/m³ – machine mixed

Description	cm 1:3		cm 1:4		gm 1:2:9	
	m^3		m^3		m^3	
Materials						
m³ cement @ £60.72/tonne						
x 1.44 tonne/m³	1	87.44	1	87.44	1	87.44
m³ lime @ £86.50/tonne						
x 0.7 tonne/m³	–		–		2	121.10
m³ sand @ £7.00/m³	3	21.00	4	28.00	9	63.00
Unloading						
(Price for cement and lime		–		–		–
includes cost of unloading)						
	÷ 4	108.44	÷ 5	115.44	÷12	271.54
	=	27.11	=	23.09	=	22.63
Allow for shrinkage	25%	6.78	25%	5.77	30%	6.79
Cost of materials/m³		33.89		28.86		29.42

Mixing
5/3¹/₂ mortar mixer/hour
Output 1 m³/hour

Hire charge	£25.40	
Mixer operator		
£4.86 x 40 hours	= £194.40	
Cleaning, etc.		
£4.86 x 5 hours	= £24.30	
Fuel	£27.20	
	─────	
	271.30	

÷ 40 x 75% = 30hrs

	cm 1:3	cm 1:4	gm 1:2:9
Hourly cost of mixing	9.04	9.04	9.04
Total cost of mortar 1 m³	42.93	37.90	38.46

Mixing mortar

Mixing mortar by machine can be cheaper than mixing by hand. However, machine mixing is suitable for projects requiring a substantial amount of mortar.

Labour cost

The element of labour cost in brickwork comprises the wages of bricklayers and labourers who work in gangs (e.g. two bricklayers and one labourer). Labourers mix mortar and transport mortar and bricks to the bricklayers' work area. The gang structure, that is, the ratio of bricklayers to labourers, depends on the following factors:

1. *Output of bricklayers* Where the output of the bricklayers is high then the gang size should be large to keep the bricklayers in constant supply of bricks and mortar (see Table 6.13).
2. *Distance between work and materials* More labourers are required to move materials to work areas where a long distance has to be covered.
3. *Method of transporting materials* The gang ratio will be small where mechanized methods of transporting material is adopted. But if wheelbarrows are used or where materials are carried by labourers using hods, then more labourers will be needed.

Table 6.13 Average number of bricks laid by a bricklayer/hour

Walls	Number of bricks	
	65 mm bricks	65 mm heavy bricks
1/2 brick (102.5 mm)	50	30
1 brick (215 mm)	55	35
1 1/2 bricks (327.5 mm)	60	40
2 bricks (430 mm)	65	45
1/2 brick skin of hollow wall	44	25

BRICK FACEWORK

Facework in brick or blockwork is any work finished fair either on one face or both faces. Work to be finished fair will be given in the description of a bill item. Table 6.14 shows the number of facing

Table 6.14 Number of facing bricks per m^2	
Type of wall bond	Number of bricks
Stretcher bond	59
Header bond	118
English bond	89
Flemish bond	79
English garden wall	74
Flemish garden wall	67

bricks (size 215 x 102.5 x 65 mm with 10 mm thick mortar joint) required per square metre of brickwork.

Bricklayers' output is reduced by work required to be finished fair because of greater care and attention required. In pricing one of the typical bill items under this section, an output of 40 number bricks per hour has been taken for work finished fair.

BLOCKWORK

Blocks are units which are larger than bricks and may be classified as follows: clay blocks (generally of extruded hollow units); foamed slag; and concrete blocks. Blocks can be lightweight, dense, solid, hollow or cellular, and can be obtained in various faces, sizes, thicknesses, strengths. However, the most common sizes are shown in Table 6.15.

Table 6.15 Size of blocks		
Length (mm)	Height (mm)	Thickness (mm)
290	215	58, 60, 75, 100, 125, 140, 190, 215
440	140	58, 60, 75, 100, 125, 140, 190, 215
440	215	58, 60, 75, 100, 125, 150, 190, 215

Unlike bricks, blocks are purchased by the square metre, and depending on the thickness of the block, average unloading output ranges from 0.06 hours per m² for 50 mm thick blocks to 0.12 hours per m² for 215 mm thick blocks.

DAMP-PROOF COURSES

Materials used in damp-proof courses are many. The most commonly used are as follows:

1. *Bituminous*   Falling into the categories of either fibre base, hessian base or lead core.
2. *Polythene*   This is a special grade of heavy sheeting used solely for damp-proofing work.
3. *Sheet Lead*   This is an expensive material which becomes economical when used with bituminous felt damp-proof course.
4. *Sheet Copper*
5. *Slate*   Two courses of slates laid breaking joint in cement mortar.
6. *Bricks*   Blue engineering bricks.

Apart from slates and blue bricks, all damp-proof courses are purchased in rolls and in various widths.

UNIT RATE BUILD-UP FOR TYPICAL BILL ITEMS

Example 6.14

65 mm common bricks; in stretcher bond; in cement mortar (1:3)   **m²**

DATA
(i)   Half-brick skin of hollow wall below damp-proof course
(ii) Labour unloading bricks – 1.2 hours/thousand bricks
(iii) Gang ratio – 2 bricklayers to 1 labourer
(iv) Basic rates:
   Portland cement   £ 60.11/tonne delivered
   Common bricks   £112.00/thousand delivered
   Sand   £ 7.00/m³ delivered
   Bricklayer   £ 5.53/hour
   Labourer   £ 4.72/hour

		£	£
SYNTHESIS			
(i)	Materials		
	1 000 Bricks @ £112/thousand delivered	112.00	
	Unload 1.2 hours @ £ 4.72/hour	5.66	
		£117.66	
	59 common bricks @ £117.66/thousand	6.94	
	Allow for waste (5%)	0.35	7.29
	Mortar		
	1 m³ cement mortar (1:3) (from Table 6.12)	42.93	
	Allow for waste (5%)	2.15	
	Cost of 1 m³ mortar	45.08	
	Therefore 0.03 m³ cement mortar @ £45.08/m³		1.35

(ii) Labour
Gang ratio 2:1 =
£5.53 + £5.53 + £4.72 = £15.78 ÷2 = £7.89/hour
59 bricks/m² @ 44 bricks/hour

C/F   8.64

B/F   8.64

Therefore $\dfrac{59}{44}$ = 1.34 hours @ £7.89/hour   10.57

Net unit rate per m²   **£19.21**

Example 6.15

Walls – thickness 215 mm; 65 mm common brick; in cement mortar (1:4). m²

DATA
(i)   Brick wall above damp-proof course
(ii)   Labour unloading bricks – 1.2 hours/thousand bricks
(iii) Gang ratio – 2 bricklayers to 1 labourer
(iv)   Basic rates (as Example 6.14)

SYNTHESIS   £   £
(i)   Materials (as Example 6.14) = £117.66

       118 number common bricks @ £117.66/thousand   13.88

       Allow for waste (5%)   0.69   14.57
                                                      _____

       Mortar
       1 m³ of cement mortar (1:4)
            (from Table 6.12)   37.90

       Allow for waste (5%)   1.90
                                                      _____
       Cost of 1 m³ mortar   39.80
                                                      ======

       Therefore 0.07 m³ cement mortar @ £39.80/m³   2.79

(ii)   Labour
       Gang ratio 2:1
       118 bricks/m² @ 55 bricks/hour
       Therefore = $\frac{118}{55}$ = 2.15 hours @ £7.89/hour
            (as Example 6.14)   16.96
                                                               _____

Net unit rate per m²   **£34.32**
                                                               ======

Example 6.16

Walls – thickness 215 mm; facing bricks Coronet Light Red Stock; £350.00 per thousand; English bond; facework both sides; in cement/lime mortar (1:2:9); flush pointing as work proceeds.
$$m^2$$

DATA
(i)   Brick wall above damp-proof course.
(ii)   Labour unloading – 2 hours/thousand bricks
(iii) Gang ratio – 2 bricklayers to 1 labourer
(iv) Labour output – 40 bricks/hour
(v)   Basic rates:
    Facing bricks   £350.00/thousand delivered
    Portland cement   £60.11/tonne delivered
    Lime   £86.50/tonne delivered
    Sand   £ 7.00/m^3 delivered
    Bricklayer   £ 5.53/hour
    Labourer   £ 4.72/hour

SYNTHESIS	£	£
(i) Materials		
1 000 facing bricks @ £350/thousand delivered	350.00	
Unload 2 hours @ £4.72/hour	9.44	
	359.44	
89 facing bricks @ £359.44/thousand	31.99	
Allow for waste (10%)	3.19	35.18
Mortar		
1 m^3 Mortar (1:2:9) (from Table 6.12)	38.46	
Allow for waste (5%)	1.92	
Cost of mortar/m^3	40.38	
Therefore 0.07 m^3 gauged mortar @ £40.38/m^3		2.83
	C/F	37.01

                                                              B/F   37.01

(ii)   Labour
        Gang ratio 2:1
        89 Bricks/m^2 @ 40 bricks/hour
        Therefore $\dfrac{89}{40}$ = 2.23 hours @ £7.89/hour
            (as Example 6.14)   17.59

Net unit rate per m^2   **£55.60**

Example 6.17

Forming cavity – 50 mm wide; between skins of hollow walls; including vertical twisted wall ties; 5 number per m² m²

DATA

(i)   Labour output – 0.10 bricklayer hour/m² and
0.05 labourer/hour both per m²

(ii) Basic rates:
Vertical twisted wall ties   £140/thousand delivered
Bricklayer   £5.53
Labourer   £4.72

	SYNTHESIS	£	£
(i)	Materials		
	1 000 wall ties @ £140/thousand delivered	140.00	
	Unload (allow)	0.75	
	Cost of 1 000 wall ties	140.75	
	Therefore 5 number wall ties @ £140.75/thousand	0.70	
	Allow for waste (5%)	0.04	0.74
(ii)	Labour		
	0.10 bricklayer hour/m² @ £5.53/hour	0.55	
	0.05 labourer hour/m² @ £4.72/hour	0.24	0.79
	Net unit rate per m²		**£1.53**

Example 6.18

Walls – thickness 100 mm; aerated concrete blocks; Thermalite 440 x 215 mm; high strength 7 blocks (7 N/mm^2) in cement mortar (1:4).   **m^2**

DATA
(i)   Blockwall above damp-proof course
(ii) Mortar required/m^2 = 0.006 m^2
(iii) Gang ratio – 2 bricklayers to 1 labourer
(iv) Labour output – 0.66 gang/hour per m^2
(v)   Basic rates:
   Blocks 100 mm thick   £7.50/m^2 delivered
   Portland cement   £60.11/tonne delivered
   Sand   £7.00/m^3 delivered
   Bricklayer   £5.53/hour
   Labourer   £4.72/hour

		£	£
SYNTHESIS			
(i)	Materials		
	1 m^2 100 mm thick blocks @ £7.50/m^2 delivered	7.50	
	Unload 0.07 hour/m^2 @ £4.72/hour	0.33	
		7.83	
	Allow for waste (5%)	0.39	8.22
	Mortar		
	1 m^3 cement mortar (1:4) (from Table 6.12)	37.90	
	Allow for waste (5%)	1.90	
	Cost of 1 m^3 mortar	39.80	
	Therefore 0.006 m^3 cement mortar @ £39.80/m^3		0.24
(ii)	Labour		
	Gang ratio 2:1		
	£ 5.53 + £5.53 + £4.72 = £15.78/hour		
	Labour output 0.66 hour/m^2 x £15.78/hour		10.41
Net unit rate per m^2			**£18.87**

Example 6.19

Damp-proof course; single layer hessian-based bitumen felt; width not exceeding 225 mm; horizontal.   m²

DATA
(i)   Damp-proof course laid on 102.5 mm thick wall.
(ii) Labour output – 0.12 bricklayer/hour and
       0.06 labourer/hour both per m²
(iii) Basic rates:
       Hessian-based bitumen felt   £4.45/m² delivered
       Craftsman   £5.53/hour
       Labourer   £4.72/hour

	SYNTHESIS	£	£
(i)	Materials		
	1 m² felt damp-proof course @ £4.45/m² delivered	4.45	
	Unload (allow)	0.05	
		4.50	
	Allow for waste (5%)	0.23	4.73
(ii)	Labour		
	0.12 bricklayer hour @ £5.53/hour	0.66	
	0.06 labourer hour @ £4.72/hour	0.28	0.94
	Net unit rate per m²		**£5.67**

CARPENTRY/TIMBER FRAMING/FIRST FIXING – SMM7, SECTION G20

Carcassing work comprises framework of roofs, floors and walls; and that for first fixings undertaken by carpenters utilizing softwood. Carcassing timber is supplied in sizes ordered and all timbers for such works are measured in linear metres.

ESTIMATING CONSIDERATIONS

Factors which influence the cost of carpentry carcassing works are as follows:

Type or species of timber

Carcassing timbers are of different species and while each vary in price, scarcity of the specified timber may add a considerable cost to the works.

Grading of timber

The type of grading used on timber classifications adds extra to the cost of timber. Timber can be visually graded but timber for structural work needs to be machine graded and is classified either as general structural (GS) or special pre-treated structural (SS).

Sawn or wrought timber

Regularized timber – wrought timber or timber sawn to a special shape – is more expensive to purchase than ordinary sawn timber. Therefore the estimator should be on the look out for timber sections of irregular shape as a lot of machining may be necessary.

Pre-treated timber

Untreated timbers are vulnerable to degradation by wood destruction insects and fungi. Therefore timber is treated for longer life. There is also the need to impregnate some timbers for increased fire retardance. These treatments are many and varied but on the average can add up to $12\frac{1}{2}$ per cent to the timber cost.

Mechanical fasteners required

The cost of supplying and fixing some of these mechanical fasteners is normally allowed for in a bill of quantities, but fixing them on site needs extra skill, experience and can be time-consuming. There is a wide range of mechanical fasteners (e.g. ragbolts, rawlbolts, special nails, steel gussets, split ring connector, toothed plate connectors, shear plate connector units). Requirement for more of these will affect the price.

Size and length of members

Timber of large cross-section and/or in long continuous lengths is expensive to purchase and awkward and difficult to fix. Where the length required cannot be obtained in one continuous length, joining with mechanical fasteners discussed above will be necessary.

Type of work

The estimator's pricing method is influenced by the type of carpentry work required. Work which requires a lot of sawing, as a result of small sections of timber needed for strutting and stiffening of members, is time-consuming. Therefore it is more expensive to produce work in smaller and shorter timber members.

Location of work

Work located in awkward positions or at high levels above ground would require a lot of temporary work. It can also be risky to carry out thus requiring additional care and time in its execution.

Purchasing timber in random or precise lengths

Timber can be purchased in random lengths for sorting or cutting into various lengths according to requirement. Whilst it is cheaper to purchase timber in random lengths, higher cost is incurred as a result of waste from cutting and labour cost in sorting, handling and cutting results. On the other hand, the initial cost of timber purchased in precise lengths is high, but waste in labour and materials, if any, is minimal.

Labour output

Table 6.16 gives a guide to labour output on carcassing work.

Table 6.16   Guide to average labour outputs per m³ for carcassing work

	Labour hours
Wall plates	18
Floor and ceiling joists	24
Roof joists	42
Purlins, ties, struts	36
Partitions	48

ROOF TRUSSES

In the current age of industrialization of the construction industry, most building materials/components arrive on site prefabricated. One of the components under this section which is often factory produced is the roof truss.

Apart from the off-site fabrication, most producers offer clients/designers a total roof design service which embraces design responsibility for the trusses, stability, bracing and *in situ* construction, bearing and fixing details. However, with the above improved services, roof trusses are sometimes measured in detail with the manufacturer's reference number quoted in the bill of quantities for pricing. The estimator may need to consider the following items of cost when pricing roof trusses:

1. *Unloading*   Although the delivery of the trusses may be made by the supplier, unloading is still the responsibility of the contractor.
2. *Extra labour*   Additional labour will be required at an intermediate position when lifting long-span trusses manually.
3. *Repair/replacement*   Damaged trusses have to be remedied or replaced at the contractor's expense.
4. *Extra care*   Trusses are fragile components and therefore require extra storage facilities, careful handling, erection and fixing.
5. *Accessibility for hoisting*   Enough space should be available for hoisting the roof trusses. In the absence of this, problems in hoisting may arise with consequential delays to the contract programme.

6. *Range of trusses* The common range of trusses produced are: fink, monopitch, bobtail and single cantilever. However, manufacturers can design/produce other ranges (e.g. triangular, dual shape, double cantilever) to suit the design requirements.

UNIT RATE BUILD-UP FOR TYPICAL BILL ITEMS

Example 6.20

Floor members; sawn softwood; 38 x 100 mm; impregnated. m

DATA
(i)   Basic rates:
      Timber   £345.50/m^3 delivered
      Nails   £ 1.80/kg
      Carpenter   £5.53/hour
      Labourer   £4.72/hour

		£	£
SYNTHESIS			
(i)	Materials		
	1 m^3 Treated timber @ £345.50/m^3 delivered	345.50	
	Unload 2 hours/m^3 @ £ 4.72/hour	9.44	
		354.94	
	Nails 1.6 kg @ £1.80/kg	2.88	
		357.82	
	Allow for waste (10%)	35.78	
		393.60	
(ii)	Labour		
	Fixing 1 m^3 timber = 24 hours/m^3 @ £5.53/hour	132.72	
	Cost of supplying and fixing 1 m^3 timber	526.32	
	Therefore cost of 1 m length of 38 x 100 mm		
	$= \dfrac{0.038 \times 0.100}{1 \text{ m}^3} \times £526.32$	2.00	

Net unit rate per m   **£2.00**

Example 6.21

Roof members; pitched; sawn softwood 25 x 125 mm; impregnated.   m

DATA
(As Example 6.20)

	SYNTHESIS	£	£
(i)	Materials		
	1 m³ treated timber @ £345.50/m³ delivered	345.50	
	Unload 2 hours/m³ @ £ 4.72/hour	9.44	
		354.94	
	Nails 1.6 kg @ £1.80/kg	2.88	
		357.82	
	Allow for waste (10%)	35.78	
		393.60	
(ii)	Labour		
	Fixing 1 m³ timber = 42 hours/m³ @ £5.53/hour	232.26	
	Cost of supplying and fixing 1 m³ timber	625.86	

Therefore cost of 1 m length of 25 x 125 mm timber

$$= \frac{0.025 \times 0.125 \text{ mm}}{1 \text{ m}^3} \times £625.86 \qquad 1.96$$

Net unit rate per m   **£1.96**

Example 6.22

Wall or partition members; 50 x 50 mm impregnated.   m

DATA
(As Example 6.20)

	SYNTHESIS	£	£
(i)	Materials		
	1 m³ treated timber @ £345.50/m³ delivered	345.50	
	Unload 2 hours/m³ @ £ 4.72/hour	9.44	
		354.94	
	Nails 1.6 kg @ £1.80/kg	2.88	
		357.82	
	Allow for waste (10%)	35.78	
		393.60	
(ii)	Labour		
	Fixing 1 m³ timber = 48 hours/m³ @ £5.53/hour	265.44	
	Cost of supplying and fixing 1 m³ timber	659.04	

Therefore cost of 1 m length of 50 x 50 mm timber

$$= \frac{0.050 \times 0.050 \text{ mm}}{1 \text{ m}^3} \times £659.04 \qquad 1.65$$

Net unit rate per m   **£1.65**

Example 6.23

Sawn softwood structural roof timber; Hunter Trussed Rafters; GS grade; impregnated; 35 degree monopitch; 450 mm overhang; 4 000 mm span over wall plates; fixing with clips (measured separately). Nr

DATA
(i) Labour output – fixing roof truss, carpenter 1.20 hours each
(ii) Basic rates:
    35 degree monopitch truss 4 000 mm span
      £23.50 each delivered
    Carpenter £5.53/hour
    Labourer £4.72/hour

SYNTHESIS	£	£
(i) Materials		
35 degree monopitch truss @ £23.50 each	23.50	
Unload 0.50 hour @ £4.72/hour	2.36	
	25.86	
Allow for waste (5%)	1.29	
	27.15	
(ii) Labour		
Hoisting: labourer 0.10 hour @ £4.72/hour = 0.47		
Fixing: carpenter 1.20 hours @ £5.53/hour = 6.64	7.11	
	34.26	

Net unit rate per Nr   **£34.26**

Example 6.24

Galvanized steel straps and clips; BAT; 30 x 2.5 mm standard strapping; 1 200 mm long; fixing with galvanized nails.   Nr

DATA
(i)   Labour output – 0.02 hour each
(ii) Basic rates:
      30 x 2.5 mm galvanized steel straps   £1.40 each delivered
      Carpenter   £5.53/hour
      Labourer   £4.72/hour

SYNTHESIS	£	£
(i) Materials		
30 x 2.5 mm galvanized steel straps £1.40 each	1.40	
Galvanized nails (allow)	0.10	
	1.50	
Allow for waste (5%)	0.08	
	1.58	
(ii) Labour		
Carpenter 0.02 hour @ £5.53/hour	0.11	
	1.69	

Net unit rate per Nr   **£1.69**

CLAY/CONCRETE AND LEAD ROOF COVERING – SMM7, SECTIONS H60, H71

Work covered in this section consists of tile roof covering of various shapes, sizes and colours. These are available in concrete or clay. This work is normally undertaken by subcontractors. However, the estimator builds up comparable unit rates which assist him or her in the evaluation and selection of the most competitive subcontractor's quotations for the tender.

GENERAL POINTS

Most roof tiles have two projecting nibs on the underside and one or two holes. Both the holes and nibs are located near the head for hanging and/or nailing. The provision of the nibs enables tiles to be fixed without nails, hence the tiles require nailing on every fourth course. However, nailing is required on every tile without nibs at abutments, verges, swept and laced valleys, and very exposed positions.

Roof tiles are purchased by the thousand. They are often fragile and therefore require a good allowance for waste during purchase. Moreover the total measured area given in the bill of quantities is nett without any allowance for laps. It is therefore essential that the formula used in calculating the required number of tiles for a given area is understood, thus enabling the incorporation of the necessary allowance for laps. The method of calculating the number of tiles required per square metre is as follows:

$1 \text{ m}^2 \div {}^1/_2$ gauge of tile x width of tile, or

$$\frac{1 \text{ m}^2}{{}^1/_2 \text{ gauge of tile x width of tile}}$$

For example:

265 x 165 mm tiles with 65 mm lap

$$\text{Gauge} = \frac{\text{length of tile} - \text{lap}}{2} = \frac{265 - 65}{2} = 100 \text{ mm}$$

$$\text{Therefore number of tiles per m}^2 = \frac{1\,000 \text{ (mm)}}{100 \text{ (mm)}} \times \frac{1\,000 \text{ (mm)}}{165 \text{ (mm)}} = 61 \text{ number}$$

SOFTWOOD BATTENS

Tiles are hung and/or nailed to softwood battens and the method of calculating the total length required per square metre of tiling is as follows:

Number of tiles per m² x width of tile

For example:

265 x 160 mm tiles with 65 mm lap

Number of tiles per m² = 61 x width of tile (165) = 10.07m

ESTIMATING CONSIDERATIONS

Factors affecting cost of tile roof coverings may include the following.

Type of tiles

The number of tiles required per given area depends on the type of tile: that is, whether it is centre- or head-nailed and whether it is a plain or interlocking tile.

Height of roof above ground level

Work to the roof of a tall building will require an expensive temporary scaffolding and protective work. In addition, extra time will be required to get tiles and operatives to the various work places and safety measures required may reduce operatives' output.

Complexity of roof

Roofs have become an essential architectural feature for most buildings and, as such, are sometimes complex in construction and finishing. The roof of a building may have varying heights, pitches and slopes, flashings, gutters and hips which require extra care and attention in its execution. This will therefore involve an additional cost in material and time consumed.

Location of site

Tiling work to buildings located in built-up areas close to the public will require extra protective items and care, and will influence the estimator's pricing level.

Time of year

Temporary protection may be required if work is undertaken during severe winter months. This can be expensive if the roof is of complex shape.

Average output

Table 6.17 is a guide to the average coverage and labour output on tiling work.

Table 6.17	Guide to average coverage and labour output for roof tiling work			
Size of plain tiles		*Number of tiles per m²*	*Length of batten per m*	*Number of tiles laid per gang hour*
Gauge (mm)	*Lap (mm)*			
265 x 165 mm				
100	65	61	10	100
95	75	64	11	100
Interlocking tiles *380 x 230 mm*				
152.5	75	10	3	50
420 x 330 mm				
172.5	75	16	3	55
Taking delivery and stacking tiles		1.75 hours per 1,000		
Fixing softwood battens		1 hour per 50 m		

SHEET METAL ROOFING, FLASHINGS AND GUTTERS

Sheet metal roofing employs materials such as lead, zinc, copper and aluminium. While these materials are durable roof coverings, their main determining factor is flat roofs. In addition lead is much heavier than tiles and slates, but because of its malleability, it is employed in work items such as flashings, aprons, valleys, gutters, and so on.

UNIT RATE BUILD-UP FOR TYPICAL BILL ITEMS

Example 6.25

Roofing; pitched 40 degrees from horizontal; plain concrete tiles; 265 x 165 mm; laid to a 65 mm lap and fixing each tile with two galvanized nails in every fourth course; 38 x 25 mm pressure impregnated softwood battens on reinforced under-slating felt (BS.747). m²

DATA
(i)   Nails   1.75 kg/1 000 tiles
(ii)   Gang ratio – 1 roofer to 1 labourer
(iii) Basic rates:
     Concrete tiles   £375.00/thousand delivered
     Timber   £155.00/m³ delivered
     Nails   £7.32/kg
     Reinforced under-slating felt   £1.10/m²
     Roofer   £5.53/hour
     Labourer   £4.72/hour

SYNTHESIS
	£	£
(i) Materials		
1 000 tiles @ £375.00/thousand delivered	375.00	
Unload and stack 1.75 hours/1 000 @ £4.72/hour	8.26	
	383.26	
Nails 1.75 kg/1 000 tiles @ £7.32 = £12.81	12.81	
	396.07	
Allow for waste (5%)	19.80	
Cost of 1 000 tiles, etc	415.87	
61 number tiles per m² (see p. 157) @ £415.87/1 000		25.37
1 m² under-slating felt @ £1.10/m²	1.10	
Allow for waste and laps (10%)	0.11	1.21
	C/F	26.58

		B/F	26.58
Softwood battens			
1 m³ timber @ £155.00/m³ delivered	155.00		
Impregnation (allow) 10%	15.50		
Unload 2 hours @ £4.72/hour	9.44		
Allow for waste (5%)	9.00		
	188.94		

Length of battens required for 1 m² tiles = 10.07 m (see p. 158)
Therefore 10.07 m x 38 mm x 25 mm = 0.0096 m³

0.0096 m³ timber @ £188.94/m³	1.81	
Nails (allow) 5%	0.09	1.90
		28.48

(ii)   Labour
Cost of gang hour

One roofer	£ 5.53
One labourer	£ 4.72
	£10.25

Fixing 1 m² felt

0.10 gang hour @ £10.25/hour	1.03	
Nails (allow) 5%	0.05	

Fixing battens

$$10.07 \text{ m @ } 50 \text{ m/hour} = \frac{10.07 \times £10.25}{50 \text{ m}}$$ 2.06

Fixing tiles

$$61 \text{ tiles @ } 100 \text{ tiles per hour} = \frac{61 \times £10.25}{100}$$ 6.25   9.39

Net unit rate per m²   **£37.87**

Example 6.26

| Double course at eaves; 215 x 165 mm plain tiling. | m |

DATA
(As Example 6.25)

SYNTHESIS   £   £
Based on 10 metre length
(i)   Materials
       Tiles (as Example 6.25)
       Number of tiles 165 mm wide in 10 m width

$$= \frac{10 \text{ m}}{165 \text{ mm}} = 61 \text{ number (see p. 157)}$$

(see p. 157)

       Therefore 61 number tiles @ £415.87/thousand   25.37
                                                            _____

       Softwood battens (additional row)
       Cost of timber/m³ (As Example 6.25) = £188.94
       10.00 x 38 mm x 25 mm 0.0095 m³
       Therefore 0.0095 m³ timber @ £188.94/m³   1.79

       Nails (allow) 5%   0.09
                                                            _____
                                                            27.25

(ii)   Labour
       Fixing batten

       $10 \text{ m @ } 50 \text{ m/hour} = \dfrac{10 \times £10.25}{50}$ 2.05

       Fixing tiles

       $61 \text{ Tiles @ } 100 \text{ tiles/hour} = \dfrac{61 \times £10.25}{100}$ 6.25
                                                            _____
       Cost of 10 m length eaves tiling   35.55
                                                            ══════

       Therefore cost of 1 m length = $\dfrac{£35.55}{10}$ 3.56

Net unit rate per m   **£3.56**
                                                                    ══════

Example 6.27

| Ridges; ridge tiles to roof tiling; half-round butt-jointed; bedding and pointing in cement mortar (1:3). | m |

DATA
(i)   0.003 m³ mortar required for 1 metre length
(ii) Labour output – unload and stack tiles 0.5 hour/100 tiles
(iii) Gang ratio – 1 roofer to 1 labourer
(iv) Basic rates:
      Ridge tiles 456 mm long   £276.00/100 delivered
      Roofer   £5.53/hour
      Labourer   £4.72/hour

SYNTHESIS

(i) Materials

	£	£
Ridge tiles 100 No 456 mm long @ £276.00/hundred delivered	276.00	
Unload and stack 0.5 hour @ £4.72/hour	2.36	
	278.36	
Allow for waste (5%)	13.92	
Cost of 100 number ridge tiles	292.28	

$$\text{Number of tiles/m} = \frac{1\,000\text{ mm}}{456\text{ mm}} = 2.193 \text{ number}$$

Therefore 2.19 number tiles @ £292.28/100 tiles		6.40
Mortar		
0.003 m³ mortar @ £42.93/ m³ (from Table 6.12)	0.13	
Allow for waste (5%)	0.01	
		0.14
(ii) Labour		
Fixing tiles		
0.15 gang hour @ £10.25/hour		1.54

Net unit rate per m   **£8.08**

Example 6.28

Flashings; Code 4 lead; 150 mm girth; 150 mm lapped joints; fixing to masonry with nailed lead clips and lead wedges.   m

DATA
(i)   Gang ratio – 1 roofer to 1 labourer
(ii) Unload lead 1 hour/tonne
(iii) Labour output – 0.25 hour/metre
(iv) Basic rates:
     Lead   £915.00/tonne delivered
     Roofer   £5.53/hour
     Labourer   £4.72/hour

SYNTHESIS	£	£
(i) Materials		
1 tonne lead @ £915.00/tonne delivered	915.00	
Unload 1 hour @ £4.72/hour	4.72	
	919.72	
Allow for waste (2$\frac{1}{2}$%)	22.99	
Cost of 1 tonne lead	942.71	

1 m^2 of Code 4 lead weights 20.41 kg

Coverage for 1 tonne = $\frac{1,000}{20.41}$ = 49 m^2

Cost of square metre lead = $\frac{£942.71}{49 \text{ m}^2}$ = £19.24

Therefore 150 mm wide lead @ £19.24/m^2	2.89	
Allow for laps, clips, etc. (7$\frac{1}{2}$%)	0.22	3.11
(ii) Labour		
0.25 gang hour @ £10.25/hour		2.56

Net unit rate per m   **£5.67**

BUILT-UP FELT ROOFING – SMM7, SECTION J41

Built-up felt roof covering is made up of two or more layers of bitumen felt bonded together on site with bitumen compound. Bitumen felt may be either fibre base for low cost, asbestos base for improved fire resistance or glass fibre base for improved resistance to decay. It is purchased in rolls of up to 20 metres long by 1 metre wide and is available in a number of finishes (e.g. saturated, sanded, mineral, natural, reinforced or venting base layer).

As the coverage area given in the bill of quantities is net, an allowance is required to be made for laps and this is dependent on the level of specification. The following is an example of lap allowance calculation for specification calling for 75 mm side and 100 mm end laps:

Area of roll:
10 metres x 1 metre = 10 square metres.

Area given in bill of quantities:
(10 m − 100 mm) x (1 m − 75 mm) = 9.90 m x 0.925 m = 9.1575 m².

Area consumed by laps:
10 m² less 9.1575 m² = 0.8425 m².

Percentage:
$$\frac{0.8425 \times 100}{9.1575} = 9.2\%$$

UNIT RATE BUILD-UP FOR TYPICAL BILL ITEM

Example 6.29

Roof covering; felt (BS.747 Type 1B) in two layers; bonded with hot bitumen bonding compound; pitch 40 degrees from horizontal; 75 mm side and 10 mm end laps; bonding to timber base with hot bitumen compound. m^2

DATA
(i)   Asbestos based roofing felt Type 1B
(ii) Gang ratio – 1 roofer to 1 labourer
(iii) Labour output:
   Primer   0.05 hour/m^2
   Felt, 2 layers   0.15 hour/m^2
(iv) Basic rates:
   Roofing felt   £19.75/roll (10 m^2) delivered
   Bitumen   £48.50/tonne delivered
   Primer   £ 4.50/litre
   Roofer   £ 5.53/hour
   Labourer   £4.72/hour

SYNTHESIS

		£	£
(i)	Materials		
	1 roll of felt (10 m^2 area) @ £19.75/roll delivered	19.75	
	Unload (allow)	0.20	
		19.95	
	Allow for laps and waste (15%)	2.99	
	Cost of 10 m^2 felt	22.94	

Therefore cost of 1 m^2 felt $= \dfrac{£22.94/m^2}{10} \times 2$ layers   4.59

Primer
Per litre; coverage 4 m^2
Therefore cost/$m^2 = \dfrac{£4.50}{4}$ 1.13

Bitumen
1.5 kg/layer @ £48.50/tonne x 2 layers   0.15

C/F   1.28

		B/F	1.28	
	Allow for waste (5%)		0.06	1.34

(ii) Labour
   Application of primer   0.05 hour
   2 Layers of felt   0.15 hour
                           0.20 hour @ £10.25/hour   2.05

Net unit rate per m²   **£7.98**

PLASTERBOARD DRY LINING – SMM7, SECTIONS K10, K31

Plasterboard dry lining is an internal wall and ceiling lining which can be used as a base for plastering or direct decorations. Many proprietary units are available in various thicknesses and sizes (see Table 6.18) and may be secured to softwood backgrounds with galvanized wire clout nails.

Table 6.18 Thickness and size of plasterboards		
Length (mm)	Width (mm)	Thickness (mm)
1 800, 2 400 and 3 000	600, 900 and 1 200	9.5, 12.7, 19.0 mm or thicker

ESTIMATING CONSIDERATIONS

Factors influencing the cost of dry lining may include the following.

Type of material

Some of the many different types of proprietary materials available possess thermal insulation properties, some are self-supporting, while others are used to contribute to the fire resistance of separating structural elements. Those chosen to perform specific functions vary in thickness and application methods, and cost more.

Nature of background

Plasterboard may be fixed either to softwood battens on wall backing, softwood ceiling members and studs, or plaster dots and dabs on walls at various centres. All the various backgrounds have different cost attractions.

Boards for direct decoration

Some plasterboards are ready for direct decoration and others may act as a base for plastering. Those for direct decoration require extra care during fixing as the subsequent paint does not conceal any damage on the board's surface.

Position of work

Work may be on walls and ceilings, in staircase areas, and in confined spaces. Work to ceilings, confined spaces and staircase areas is time consuming and hence attracts higher prices than work to other areas.

UNIT RATE BUILD-UP FOR TYPICAL BILL ITEM

Example 6.30

12.7 mm wall board; for direct decoration; to wall 2.1-2.4 m high; 3 mm joints filled with plaster and scrimmed; fixing with galvanized nails to timber base; over 300 mm wide.   m²

DATA
(i) Plasterboard for direct decoration
(ii) Gang ratio – 1 plasterboard fixer to 1 labourer
(iii) Average labour output – fixing 0.25 gang hour/m²
(iv) Basic rates:
    Plasterboard 1800 x 1200 mm @ £6.75/sheet delivered
    Board finish plaster   £78.50/tonne delivered
    Jute scrim cloth   £4.50/roll (100 m)
    Nails   £1.70/kg
    Plasterboard fixer   £5.53/hour
    Labourer   £4.72/hour

SYNTHESIS

(i) Materials

	£	£
1 Sheet of plasterboard @ £6.75/sheet delivered	6.75	
Unload 0.10 hour @ £4.72/hour	0.47	
	7.22	
Nails 0.10 kg @ £1.70/kg	0.17	
Scrim cloth 1.80 + 1.20 m = 3 m length per board		
$= \dfrac{£4.50 \times 3\,m}{100}$	0.14	
	7.53	
Allow for waste (5%)	0.38	7.91
1 tonne board finish plaster @ £78.50/tonne delivered	78.50	
Unload 1 hour @ £4.72/hour	4.72	
	83.22	
Allow for waste (5%)	4.16	
Cost of 1 tonne finish plaster	87.38	C/F 7.91

B/F   7.91

Allow 25 kg per 30 metres joint
Therefore cost per board (i.e. 3 m length)

$$= \frac{£87.38 \times 25 \times 3}{1\,000 \times 30}$$

0.22

8.13

Cost of materials for 2.16 m² (i.e. 1.80 x 1.20 m)
   = £8.13

Therefore cost of 1 m² $= \dfrac{£8.13}{2.16 \text{ m}^2}$

3.76

(ii)   Labour
     Fixing and filling joints of plasterboard
       0.25 gang hour @ £10.25/hour

2.56

6.32

Therefore for 2.40 m (maximum) height wall
   = £6.32 x 2.40 = £15.17

Net unit rate per m   **£15.17**

TIMBER FLOORING/SHEATHING/LINING – SMM7, SECTIONS K11, K20 AND K21

Constructed timber floors of either tongued and grooved timber boards, ordinary floor boards or sheet flooring on softwood joists supported on walls are examples considered under this section. Cost implications of the material which is basically timber has been discussed above (see pp. 148–50).

ESTIMATING CONSIDERATIONS

Factors that influence cost may include the following.

Type of work specified

The flooring specified may be either ordinary floor board, plywood sheeting with tongued and grooved joints or tongued and grooved timber boards. Tongued and grooved flooring requires materials of good quality and good workmanship. Labour and material content in each of these flooring types vary and so do their cost implications.

Grade of sheet of material

Plywood sheeting for flooring may be of several grades in accordance with the preservative treatment applied to the board. Hence plywood sheeting may be described as ordinary plywood, water resistant (WR) plywood, or water and boil proof (WBP) plywood. Each of the above plywoods attracts a different price and hence influences the unit rate build-up.

Intended use of floor

The intended use determines whether the timber floor is to be decorative (e.g. ballroom and gymnasium) or to be subsequently covered as floors in residential buildings. While the former requires expensive quarter-sawn boards for better wear and appearance, the latter requires less expensive plain sawn timber.

Thickness of material

The thickness of floor boarding can range from 19, 25, 32, 40 and 50 mm and the greater the thickness, the higher the cost of material and fixing.

Complexity of work

Timber flooring in irregular-shaped areas or an obstructive location can be costly in terms of labour time and material wastage. Work in these locations requires more labour time in cutting to shape and fitting around the obstructions.

Allowances for tongued and grooved work

In tongued and grooved timber flooring, extra timber is required in the formation of tongued and grooved joints. This results in an average loss of 13 mm per board width. For instance, the joints reduce the surface 100 mm-wide board to 87 mm when jointed. This allowance may be expressed in the percentage shown in Table 6.19.

Table 6.19   Allowance for tongued and grooved work

Board width	Allowance (%)
75 mm	17.5
100 mm	13.0
125 mm	10.5
150 mm	8.5

UNIT RATE BUILD-UP FOR TYPICAL BILL ITEMS

Example 6.31

> Softwood; sawn boarding to floors; 25 mm thick; square edges;
> 125 mm wide boards; width over 300 mm.   m²

DATA
(i) Labour output – 0.66 hour/m³
(ii) Basic rates:
     Timber   £195.00/m³ delivered
     Nails £1.75/kg
     Carpenter £5.53/hour
     Labourer £4.72/hour

SYNTHESIS

		£	£
(i)	Materials		
	1 m³ timber @ £195/m³ delivered	195.00	
	Unload 2 hours @ £4.72/hour	9.44	
		204.44	
	Nails (allow) 5 kg/m³ @ £1.75/kg	8.75	
		213.19	
	Allow for waste (7¹/₂%)	15.99	
	Cost of materials/m³	229.18	

Therefore cost of materials/m²
   = 1.0 x 1.0 x 0.025 m
   = 0.025 m³ x £229.18   5.73

(ii) Labour
     Fixing 1 m² floorboard = 0.66 hour @ £5.53/hour   3.65

                                                                    9.38

Net unit rate per m²   **£9.38**

Example 6.32

Boarding to floors, tongued and grooved joints; 25 mm thick; 125 mm wide boards; fixing to softwood joists; width over 300 mm.   m^2

DATA

Basic rates:
  Timber   £290/m^3 delivered
  Nails   £1.75/kg
  Joiner   £5.53/hour
  Labourer   £4.72/hour

	SYNTHESIS	£	£
(i)	Materials		
	1 m^3 timber @ £290/m^3 delivered	290.00	
	Unload 2 hours @ £4.72/hour	9.44	
		299.44	
	Allow for losses of width		
	for tongues (10.5%)	31.44	
	Nails (allow) 5 kg/m^3 @ £1.75/kg	8.75	
		339.63	
	Allow for waste (7^1/$_2$%)	25.47	
	Cost of materials/m^3	365.10	
	Therefore cost of materials/m^2		
	=1.0 x 1.0 x 0.025 m		
	= 0.025 m^3 x £365.10/m^3		9.13
(ii)	Labour		
	Fixing 1 m^2 floorboard 0.66 hour @ £5.53/hour		3.65
	Net unit rate per m^2		**£12.78**

Example 6.33

18 mm thick Douglas fir plywood; WBP bonded; tongued and grooved joints; to floors on timber base; width over 300 mm.

m^2

DATA

(i)   Labour output:
     Unloading   0.10 hour/sheet
     Fixing   0.33 hour/m^2

(ii) Basic rates:
     18 mm plywood (2.24 x 1.22 m) @ £13.25 delivered
     Nails   £1.75/kg
     Joiner   £5.53/hour
     Labourer   £4.72/hour

SYNTHESIS

		£	£
(i)	Materials		
	1 sheet plywood @ £13.25/sheet delivered	13.25	
	Unload 0.10 hour @ £4.72/hour	0.47	
		13.72	
	Nails (allow) 0.75 kg/sheet @ £1.75/kg	1.31	
		15.03	
	Allow for waste (5%)	0.75	
	Cost of 2.73 m^2 (i.e. 2.24 x 1.22 m) plywood	15.78	
	Therefore cost/$m^2 = \dfrac{£15.78}{2.73\ m^2}$		5.78
(ii)	Labour		
	Fixing 1 m^2 plywood 0.33 hour @ £5.53/hour		1.82
Net unit rate per m^2			**£7.60**

WINDOWS/DOORS/STAIRS – SMM7, SECTIONS L10, L20 AND L30

Work considered under this section consists of general joinery items (e.g. timber windows, doors, frames) which, due to the need for a good finish, require more care in working the timbers.

ESTIMATING CONSIDERATIONS

The following need to be considered when pricing joinery.

Nominal or finished sizes

Generally, all sizes of timber members included in the bill of quantities, unless stated as finished size, are nominal sizes. Nominal size timber is without an allowance for planing waste (normally 3.20 mm off each sawn face) and therefore reduces in size when planed for fixing. Conversely, finished size timber has an allowance for planing waste when bought and therefore maintains its size after planing.

Table 6.20 shows that finished size timber 100 x 125 mm requires the purchase of a non-standard size timber 106.4 x 131.4 mm which will attract extra costs.

Table 6.20   Nominal and finished timber sizes

Classification	Timber specified	Timber purchased	Timber fixed
Nominal size	100 x 125 mm	100 x 125 mm	93.6 x 118.6 mm
Finished size	100 x 125 mm	106.4 x 131.4 m	100 x 125 mm

Cross-tonguing

Cross-tonguing involves grooving, inserting, glueing and cramping plywood, hardboard or hardwood tongues. This is a time-consuming operation and an adequate allowance should be made for this operation.

Glue, glasspaper, wedges, and other sundries

These must be allowed for in all work which requires framing, jointing, morticing, and so on.

Small section members

Small section members (e.g. beads, fillets) are easily broken on site during fixing and require an increased wastage factor.

Grain matching

Grain matching of timber or sheet can affect wastage and an allowance is needed to cover this cost.

Timber lengths

Timber lengths specified where they do not accord with standard lengths available will have to be made, cut specially, or purchased and cut, which will affect wastage.

Specials

Where skirtings, architraves and similar items that are specified are not obtainable in standard sizes and/or shapes as stock items, they need to be made specially for the project and will cost more.

Cut sizes from sheet materials

Cut sizes (e.g. from plywood special veneers) may give rise to high wastage as the surplus material may not be usable.

Hardness of species

Difficulties encountered in working on various species of timber are not the same, and Table 6.21 is a guide to labour multipliers which relate to some species of timber.

Table 6.21 Labour multipliers	
Types of timber	*Labour multipliers*
European redwood	1
Obeche	$1^{1}/_{2}$
Ash, beech, birch, oak, sapele, mahogany, sycamore	2
Afromosia, gabon, jarrah, teak	3

Labour output

Table 6.22 is a guide to average labour output.

Ready-made joinery items

In order to reduce construction cost and also as a result of the industrialization of the construction process, currently standard ready-made joinery items are mostly specified by architects. Therefore, prefabricated doors, windows, staircases, and so on, can be purchased from suppliers for fixing into prepared openings, thereby reducing the amount of machining and manual labour required on these components.

Table 6.22 Guide to average labour output for joinery

Item	Hours per item	
	Joiner	*Labourer*
Hanging flush doors		
686 x 1981 x 35 mm	1.00	0.10
762 x 1981 x 35 mm	1.10	0.12
813 x 2040 x 35 mm	1.25	0.13
838 x 2040 x 35 mm	1.35	0.15
686 x 1981 x 44 mm	1.15	0.12
762 x 1981 x 44 mm	1.25	0.14
813 x 2040 x 44 mm	1.35	0.16
838 x 2040 x 44 mm	1.45	0.18
Fixing casement windows		
631 x 1050 mm	1.10	0.55
915 x 1050 mm	1.20	0.60
915 x 1200 mm	1.25	0.65
1200 x 900 mm	1.25	0.65
1200 x 1200 mm	1.40	0.70
1524 x 1050 mm	1.50	0.75

UNIT RATE BUILD-UP FOR TYPICAL BILL ITEMS

Example 6.34

> Casement window in wrought softwood; 631 x 1 050 mm; wrought softwood cills; hinges; fasteners; securing to masonry with galvanized steel cramps – 4 number; factory, knotting and priming. **Nr**

DATA

(i) Basic rates:
631 x 1 050 mm casement window @ £47.00 each delivered
Joiner   £5.53/hour
Labourer   £4.72/hour

SYNTHESIS	£	£
(i) Materials		
1 number casement window (631 x 1 050 mm) @ £47.00 each delivered	47.00	
Unload and store (allow)	0.50	
	47.50	
Screws and plugs (allow)	0.75	
	48.25	
Allow for waste (2$\frac{1}{2}$%)	1.21	
	49.46	
(ii) Labour		
Joiner 1.10 hour @ £5.53 = £5.89		
Labourer 0.55 hour @ £4.72 = £2.60		
	8.49	
	57.95	

Net unit rate per Nr **£57.95**

Example 6.35

Flush door; 838 x 198 x 35 mm thick internal quality solid core; 6 mm thick plywood faced; hardwood lipping to edges.   Nr

DATA
(i)   Labour output – hanging door 1 hour (joiner) 0.10 hour (labourer)
(ii)   Basic rates:
   35 mm thick flush door 838 x 1 981 mm @ £26.50
      each delivered
   Joiner   £5.53/hour
   Labourer   £4.72/hour

	SYNTHESIS	£	£
(i)	Materials		
	1 number door @ £26.50 each delivered	26.50	
	Unload and store (allow)	0.50	
		27.00	
	Allowance to cover damage, etc. $(2^1/_2\%)$	0.68	27.68
(ii)	Labour		
	Fitting and hanging door		
	Joiner 1 hour @ £5.53/hour	5.53	
	Labourer 0.10 hour @ £4.72/hour	0.47	6.00
	Net unit rate per Nr		**£33.68**

Example 6.36

Set; 38 x 113 mm; jambs rebate twice; plugging at 300 mm centres; screwing. m

DATA
(i)   Labour output – fixing softwood lining 0.33 hour/m
(ii)   Basic rates:
     38 x 113 mm softwood lining @ £4.55/m delivered
     Joiner   £5.53/hour
     Labourer   £4.72/hour

SYNTHESIS

		£	£
(i)	Materials		
	1 m softwood lining @ £4.55/m delivered	4.55	
	Unload (allow)	0.05	
		4.60	
	Screws, fixings, etc. (allow)	0.10	
		4.70	
	Allow for waste (2^1/$_2$%)	0.12	
			4.82
(ii)	Labour		
	Fixing 1 m lining 0.33 hour @ £5.53/hour		1.82
Net unit rate per m			**£6.64**

Example 6.37

Straight flight standard staircase; wrought softwood; 25 mm treads; 19 mm risers; glued, wedged and blocked, 32 mm hardwood handrail; two 32 x 140 mm balustrade knee rails; 100 x 50 mm joist support 32 x 50 mm stiffeners; 150 x 100 mm newel posts with hardwood newel caps 864 mm wide; 2 680 mm going; 2 600 mm rise overall.   Nr

DATA
(i) Labour output:
        Joiner   13 hours
        Labourer   1.25 hours
(ii) Basic rates:
        Staircase 864 mm wide x 2 600 mm rise @ £400.00
            each delivered
        Joiner   £5.53/hour
        Labourer   £4.72/hour

		£	£
SYNTHESIS			
(i)	Materials		
	Staircase 864 mm wide x 2 600 mm rise		
	@ £400.00 each delivered	400.00	
	Unload and store 0.25 hour @ £4.72/hour	1.18	
		401.18	
	Nails, screws, plugs and sundry (allow)	0.50	
		402.68	
	Allow for waste ($2^1/_2$%)	10.07	
		412.75	
(ii)	Labour		
	Joiner 13 hours @ £5.53/hour = £71.79		
	Labourer 1.25 hours @ £4.72/hour = £ 5.90		
		77.79	
		490.54	

Net unit rate per Nr   **£490.54**

GENERAL GLAZING – SMM7, SECTION L40

Glazing is the fixing of glass or similar material in framework or an opening. Glass may be glazed with putty, beads, gaskets, and a combination of gasket and glazing compound.

TYPES OF GLASS

The different types of glass available may be considered under the following headings.

Translucent glass

This type of glass can be either rough cast, wired rough cast or patterned glass. The most common thickness manufactured is 6 mm.

Transparent glass

Under this heading are the following types:

1. Clear sheet glass which can be glazing of ordinary quality (OQ), selected quality (SQ), or special selected quality (SSQ), and is manufactured in thicknesses of from 2 to 6 mm.
2. Clear plate glass and polished plate glass, which are available in 6 to 12 mm thicknesses.
3. Float glass which can be in glazing quality for glazing (GG), selected glazing quality (SG) or silvering quality (SQ), and is manufactured in thicknesses of from 3 to 12 mm.
4. Special glass (e.g. toughened glass, armour plate) manufactured in thicknesses of 10 to 12 mm.

ESTIMATING CONSIDERATIONS

In recent times, most components arrive on site pre-glazed and cost of glass therefore forms part of the total component cost. Accordingly, besides the odd partition, door or window which may require glass and hence needs pricing, the estimator tends not to be concerned with pricing the specialist glazing work. Instead he or she adapts the unit rates of the most competitive subcontractor's quotation for the tender.

FACTORS AFFECTING COST

The price of glass and glazing can vary considerably depending on quality, size, use, location, and so on. The following factors may influence the estimator's level of pricing.

Type of glass

As indicated above, there are many types of glass possessing various properties and ranging from sheet glass to solar reflective, anti-bullet toughened glass, anti-solar glass, and heat-resisting glass, all in various thicknesses. Each of these have different costs which influences the estimator's pricing levels.

Shape of glass

Most glass is rectangular or square in shape. However, it can be irregular in shape (e.g. circular, octagonal) for specific applications. These shapes are reflected in the price of the glass as the wastage factor increases.

Size of glass

Glass that is extra large in size or extra heavy will require additional labour to lift it into the required glazing position, and hence attracts additional cost.

Number of panes in an area

The number of panes of glass in a given area determine the amount of labour and materials required in executing the glazing work. The smaller the pane and the larger the number of panes in a given area, the greater the time required to cut and glaze and hence the greater the cost per square metre.

Location of glass

Glazing work located in an awkward obstructive position or in a position with accessibility problems (e.g. working off ladders and scaffolding) is time-consuming and hence attracts a high price.

Level of protection required

The level of protection required for the finished work is determined by the location of the site and the level of known vandalism in the area. Where vandalism is prevalent, the finished glazed work will require protection until the project is handed over to the client. This will influence the estimator's approach to pricing.

Method of glazing

The simplest form of glazing takes the form of fixing glass into position using putty and sprigs, softwood beads, clips, and so on. Some form of glazing requires expensive protected metal beads and bars which must be considered in the build-up of the unit rates.

Single or multiple glazing

Multiple glazing requires double the materials and extra labour time in carrying out the work. Although most double glazing is factory-made, the extra cost is reflected in its prices.

Glazing background

Glass and glazing to metal sashes and stone needs extra material and care. They increase labour time and hence an allowance should be made for this in the unit rate build-up.

Labour output

Owing to the variety of glass types, glazing backgrounds and locations, an average labour output is difficult to arrive at since every operation has its own pecularities or special conditions. However, Tables 6.23 and 6.24 give some guidance on glazing clear sheets of glass to softwood.

Table 6.23 Guide to average labour outputs for glazing

	Thickness/hr	
Size of pane	2–4 mm	5–6 mm
Not exceeding 0.15 m²	0.80	0.90
0.15–4.00 m²	0.50	0.60

Table 6.24   Amount of putty required per m² of glazing

Size of pane	Wood sashes
Not exceeding 0.15 m²	3.0 kg
0.15–4.00 m²	0.75 kg

Softwood beads

The price for glass glazed with softwood beads does not include the cost of the beads, which is normally measured separately in the bill of quantities.

UNIT RATE BUILD-UP FOR TYPICAL BILL ITEMS

Example 6.38

Standard plain glass; clear float 3 mm thick; to wood rebates with putty; in panes not exceeding 0.15 m²	m²

DATA
(i)   Glass cut on site
(ii) Basic rates:
       3 mm float glass   £16.50/m² delivered
       25 kg linseed oil putty   £12.00/tin
       Glazier   £ 5.53/hour

SYNTHESIS

		£	£
(i)	Materials		
	1 m² float glass @ £16.50/m² delivered	16.50	
	Unload (allow)	0.50	
		17.00	
	Sprigs (allow)	0.05	
		17.05	
	Allow for waste (5%)	0.85	17.90
	Putty 3 kg @ £12.00/25 kg	1.44	
	Allow for waste (2¹/₂%)	0.04	1.48
(ii)	Labour		
	Cutting glass 0.30 hour		
	Glazing to wood 0.80 hour		
	1.10 hours @ £5.53/hour		6.08

Net unit rate per m²   **£25.46**

Example 6.39

Cast and patterned glass; rough cast; 6 mm thick; to wood rebates with wood beads (included elsewhere) and putty; in panes 0.15–4.00 m^2 m^2

DATA
(i) Glass purchased already cut
(ii) Basic rates:
     5 mm tinted patterned glass   £32.25/m^2 delivered
     25 kg linseed oil putty   £12.00/tin
     Glazier   £ 5.53/hour

SYNTHESIS	£	£
(i) Materials		
1 m^2 patterned glass @ £32.25/m^2 delivered	32.25	
Unload (allow)	0.55	
	32.80	
Sprigs (allow)	0.05	
	32.85	
Allow for waste (10%)	3.29	36.14
Putty 0.75 kg @ £12.00/25 kg	0.36	
Allow for waste (2^1/$_2$%)	0.01	0.37
(ii) Labour		
0.50 hour @ £5.53/hour		2.77
Net unit rate per m^2		**£39.28**

SAND CEMENT/CONCRETE AND OTHER FLOOR SCREEDING – SMM7, SECTION M10

Work considered under this section consists of floor screeds which generally provide a suitable bed to flooring without the need for a levelling compound.

TYPES OF FLOOR SCREED

Screeds are usually laid on a concrete base and are of the following types: concrete screed; dense aggregate cement screeds; modified cement and sand screeds; lightweight concrete screeds; and synthetic anhydrite screeds.

Materials for these floor screeds vary from Portland cement and aggregates to proprietary screeds that employ various chemical compounds to achieve a specified design.

ESTIMATING CONSIDERATIONS

Unit rate build-up for floor screeds is similar in principle to that for cement and lime mortar in masonry work.

FACTORS INFLUENCING COST

Factors influencing the cost of floor screeds may include the following.

Thickness of screed

Depending on the function of the floor screed, thickness ranges from 12.5 mm to 125 mm or more. The thicker the floor screed, the more the material and labour hours required for every square metre laid.

Number of obstructions

All forms of obstructions slow down floor screeding work. The kind of obstructions that may be encountered in floor screeding are working around pipes and outlets; working to channels or forming channels in screed; and working screed over floor heating pipes and equipment.

Evenness of base

Irregularities on the base and slopes require substantially more material and labour hours to even it out and this cost should be reflected in the pricing.

Type of finish required

Depending on the surfacing applied to them, screeds may require a trowelled, floated or screeded finish. Trowelled finish requires more labour hours.

Constituents of mix

The mix of ordinary concrete screed is one part of cement to one, two, three or four parts of sand. However, the cost increases when other materials are incorporated (e.g. granite chips, carborundum powder and chemical compounds).

Location of work

Where work is situated in an awkward position with difficult access or executed in narrow widths, the cost increases as labour output diminishes.

Work in a special pattern

Where the contract condition places obligations on the contractor in a form of an intricate pattern or method of executing the work, the cost of this should be reflected in the pricing.

Labour output

Table 6.25 gives a guide to average labour output for screeding work.

Table 6.25   Guide to average labour output for screeding

Screed thickness	Screed hours/m²		
	Screeded	Floated	Trowelled
13–19 mm	0.20	0.25	0.27
25–32 mm	0.25	0.30	0.31
38–52 mm	0.30	0.35	0.36

Note: The gang ratio is one craftsman to one labourer

UNIT RATE BUILD-UP FOR TYPICAL BILL ITEMS

Example 6.40

25 mm thick cement sand (1:4) screeded bed; to floors on concrete base; level and to falls only not exceeding 15 degrees from horizontal   m²

DATA
(i)   Gang ratio – 1 craftsman to 1 labourer
(ii) $5/3^1/2$ mixer output – 1 m³/hour
(iii) Basic rates:
    Portland cement   £56.00/tonne delivered
    Sand   £6.00/m³ delivered
    Hire of $5/3^1/2$ mixer including fuel
      and oil   £2.57/hour
    Mixer operator   £4.86/hour
    Pavior   £5.53/hour
    Labourer   £4.72/hour

SYNTHESIS

(i) Materials

	£	£
1 tonne bagged cement @ £56/tonne delivered	56.00	
Unload 1 hour @ £4.72/hour	4.72	
	60.72	
1 m³ cement @ £60.72/tonne x 1.442 tonne/m³	87.56	
4 m³ sand @ £ 6.00/m³	24.00	
	111.56	
Voids (allow) 25%	27.89	
Cost of 5 m³ cement mortar	139.45	

Therefore 1 m³ = $\dfrac{£139.45}{5}$ 27.89

Cost of mixing 1 m³ cement mortar:
Hire of $5/3^1/2$ mixer @ £2.57/hour = £2.57
Mixer operator @ £4.86/hour   = £4.86
        7.43

C/F 35.32

		B/F	35.32	
Allow for compression and waste (10%)			3.53	
Cost of cement mortar per m^3			38.85	
Therefore cost/m^2 of 25 mm thick cement mortar = £38.85 x 0.025			0.97	
Screed laths (allow)			0.30	1.27
(ii)	Labour – laying 0.25 gang hour/m^2 @ £10.25/hour			2.56
Net unit rate per m^2				**£3.83**

Example 6.41

32 mm thick granolithic pavings (1:2$^{1}/_{2}$) to floor on concrete base; level and to falls only not exceeding 15 degrees from horizontal. **m^2**

DATA
(i)   Gang ratio – 1 craftsman to 1 labourer
(ii) 5/3$^{1}/_{2}$ mixer output – 1 m^3/hour
(iii) Basic rates:
   Granite chippings   £23.75/tonne delivered
   Portland cement   £56.00/tonne delivered
   Hire of 5/3$^{1}/_{2}$ mixer including fuel
      and oil   £2.57/hour
   Mixer operator   £4.86/hour
   Pavior   £5.53/hour
   Labourer   £4.72/hour

SYNTHESIS

		£	£
(i)	*Materials*		
	1 tonne bagged cement @ £56/tonne delivered	56.00	
	Unload 1 hour @ £4.72/hour	4.72	
		60.72	
	1 m^3 cement @ £60.72/tonne x 1.442 tonne/m^3	87.56	
	2$^{1}/_{2}$ m^3 granite chippings @ £23.75/tonne x 1.35 tonne/m^3	80.16	
		167.72	
	Voids (allow) 30%	50.32	
	Cost of 3$^{1}/_{2}$ m^3 of mix	218.04	

Therefore cost of 1 m^3 = $\dfrac{£218.04}{3^{1}/_{2}}$ 62.30

Cost of mixing 1 m^3 mortar:
Hire of 5/3$^{1}/_{2}$ mixer @ £2.57/hour   = £2.57
Hire of mixer operator £4.86/hour   = £4.86

                                                           7.43

                                          C/F   69.73

	B/F 69.73	
Allow for compression and waste (10%)	6.97	
Cost per m³	76.70	
Therefore cost/m², 32 mm thick = £76.70 x 0.032	2.45	
Screed laths (allow)	0.30	2.75

(ii) Labour

Laying	0.40 gang hour	
Trowelling surface	0.27 gang hour	
	0.67 hour @ £10.25/hour	6.87

Net unit rate per m²   **£9.62**

PLASTERED/RENDERED/ROUGHCAST COATING – SMM7, SECTION M20

Work considered under this section comprises internal wall and ceiling finishes of jointless easily decorated surfaces, and textured weather resistance on external wall surfaces. Materials used for plastering consist of a mixture of binding materials, fine aggregates and water. The binding materials are of three main groups:

1. *Lime*   High calcium limes and semi-hydraulic limes.
2. *Gypsum*   Semi-hydrate gypsum-plasters.
3. *Portland cement*

Each of these materials can be used on its own as a plaster binding material; in practice they may be combined to provide either early strength (e.g. lime gauged with Portland cement) or improve workability (e.g. lime mixed with gypsum or Portland cement). However, various pre-mixed plasters are available on the market.

ESTIMATING CONSIDERATIONS

Factors influencing pricing may include the following.

Backgrounds

The properties or character of the background to which plaster is to be applied can affect material and labour cost of plastering as follows:

1. *Types of backgrounds*   The types of background for plastering are many (e.g. concrete, brick, block, glazed tiles, metal lathing). Each of these backgrounds has its particular difficulties for plaster application. For example, plasterwork to a non-rigid background such as metal lathing can create the problems of drying shrinkage, cracking or wavy appearance, which may require more labour and materials to rectify.
2. *Variations in suction*   Backgrounds that are susceptible to considerable variations in suction affect and retard the application of plaster work by quickly absorbing the water in plaster mix, that makes the plaster workable. It also causes plaster to shrink on drying which results in cracks, loss of adhesion to background, and so on, and may require more than two coats of plaster to get it right.

Location of work

Work located in awkward positions with accessibility problems or working off ladders costs more in labour time. Work to very high ceilings may require expensive temporary work (such as mobile or birdcage scaffold) and extra labour time in its execution.

Allowance for key waste

Thickness of plaster stated in the bill of quantities is measured from wall face. Extra cost is incurred in providing material for a key and therefore an additional material 3 mm thick is allowed as key waste to brick walls in the unit rate build-up. For metal lathing an allowance of 5 mm in material thickness will be adequate to cover the extra cost.

Allowance for compression and surface irregularities

As the base (i.e. wall surface) will have some irregularities and the plaster will also be compressed on application, an allowance should be made for the extra material involved.

Labour output

Table 6.26 gives a guide to average labour output in this work.

Table 6.26   Guide to average labour output for plastering

Operation	Gang output hrs/m²
Rendering to walls	0.25
Floating coats to walls	0.25
Setting coat to wall	0.30
Setting to soffits	0.35
Render and set	0.50
Render, float and set	0.66

Multipliers

The following multipliers in Table 6.27 should be used in variations to the operations described in Table 6.26 to take account of working on scaffolding, lifting materials and working at great heights.

Table 6.27   Multipliers for various work areas

Description	Multiplier
Rendering to walls	1.00
Work to ceilings or walls (0–3 m high)	1.10
Work to ceilings or walls (3–4.5 m high)	1.20
Work to ceilings or walls (4.5–6 m high)	(0.10 for 1.5 m increments)
External work	1.10
Circular work	2.00
Work to panels	5.00

Table 6.28   Approximate covering capacity

Mix used neat	Thickness (mm)	Approximate coverage per tonne	Type of coat
Carlite Browning	11	140 m^2	Floating
Metal Lathing	8 (from face of lath)	600 m^2	–
Bonding			
concrete	8	150 m^2	Floating
masonry	11	105 m^2	Floating
Finish	2	450 m^2	Finishing
Thistle board finish	5	165 m^2	Finishing (one coat work)

Materials coverage

Table 6.28 gives the approximate covering capacity of a range of coating materials.

PLASTERBOARD

Gypsum plasterboard is composed of a layer of gypsum sandwiched and firmly bonded between two heavy lining papers. Plasterboard is a low cost material and hence it is commonly used in internal wall and ceiling lining which can either be plastered or decorated direct. The types of plasterboard available include:

1.   Gypsum baseboard and gypsum lath for plastering.
2.   Gypsum plank for both plastering and direct decorative finish.
3.   Gypsum wall board for direct decorative finish.

For information on thickness, sizes and estimating considerations, see pp. 169–72.

UNIT RATE BUILD-UP FOR TYPICAL BILL ITEMS

Example 6.42

13 mm thick plaster; Carlite; premixed; floating coat of browning 11 mm thick; finishing coat of finish plaster 2 mm thick to
walls on brickwork or blockwork base; steel trowelled; width
over 300 mm.   m²

DATA

(i)   Coverage of browning plaster/tonne = 140 m²
(ii)   Coverage of finishing plaster/tonne   = 450 m²
(iii) Gang ratio – 2 plasterers to 1 labourer
(iv)   Hand mixing 1 m² plaster = 0.10 labourer hour
(v)   Basic rates:
       Browning plaster   £102.75/tonne delivered
       Finishing plaster   £ 82.15/tonne delivered
       Plasterer   £ 5.53/hour
       Labourer   £ 4.72/hour

SYNTHESIS   £   £

(i)   Materials
       1 tonne browning plaster @ £102.75/tonne
          delivered   102.75
       Unload 1 hour @ £4.72/hour   4.72
                                                    ――――――
                                                    107.47
       Allow for waste (5%)   5.37
                                                    ――――――
       Cost of 140 m² browning   112.84
                                                    ――――――

       Therefore cost of browning per m² = $\dfrac{£112.84}{140}$ 0.81

       1 tonne finishing @ £82.15/tonne delivered   82.15
       Unload 1 hour @ £4.72/hour   4.72
                                                    ――――――
                                                     86.87

       Allow for waste (5%)   4.34
                                                    ――――――
       Cost of 450 m² finishing plaster   91.21
                                                    ══════

                                                         C/F   0.81

B/F   0.81

Therefore cost of finishing plaster/m^2 = $\dfrac{£91.21}{450}$ 0.20

1.01

(ii)   Labour
Hand mixing 1 m^2 plaster = 0.10 hour @
£4.72/hour   =   0.47
Applying 2 coats of plaster to wall
= 0.33 gang hour @ £15.78/hour   =   5.21

5.68

Net unit rate per m^2   **£6.69**

Example 6.43

13 mm thick plaster; first and finishing coats cement and sand (1:3); wood floated onto walls on brickwork or blockwork base; widths over 300 mm. m^2

DATA
(i) Gang ratio – 2 plasterers to 1 labourer
(ii) $5/3^1/_2$ mixer output – 1 m^3/hour
(iii) Basic rates:
       Portland cement   £56.00/tonne delivered
       Sand   £6.00/m^3 delivered
       Hire of $5/3^1/_2$ mixer (including fuel and oil)   £2.57/hour
       Plasterer   £5.53/hour
       Mixer operator   £4.86/hour

SYNTHESIS

		£	£
(i)	Materials		
	1 tonne bagged cement @ £56/tonne delivered	56.00	
	Unload 1 hour @ £4.72/hour	4.72	
		60.72	
	1 m^3 cement @ £60.72/tonne x 1.442 tonne/m^3	87.56	
	3 m^3 sand @ £ 6.00/m^3	18.00	
		105.56	
	Voids (allow) 25%	26.39	
	Cost of 4 m^3 cement mortar	131.95	
	Therefore cost of 1 $m^3 = \dfrac{£131.95}{4}$	32.99	
	Cost of mixing 1 m^3 cement mortar:		
	Hire $5/3^1/_2$ mixer @ £2.57/hour = £2.57		
	Mixer operator @ £4.86/hour = £4.86		
		7.43	
		40.42	
	Allow for compression and waste (10%)	4.04	
	Cost of 1 m^3 cement mortar	44.46	

   Therefore cost per m², 16 mm (13 + 3) thick
      = £44.46 x 0.016   0.71

(ii) Labour
      Applying plaster = 0.25 gang hour @ £15.78/hour   3.95

                                                                        _____
Net unit rate per m²   **£4.66**
                                                                        =======

Example 6.44

Baseboarding; 19 mm thick Gyproc plank; square edge; 5 mm joints filled with plaster and scrimmed; work to ceilings; width over 300 mm; fixing with galvanised nails to timber base.   m²

DATA
(i)   Plasterboard for direct decoration
(ii) Gang ratio – 1 plasterboard fixer to 1 labourer
(iii) Average gang labour output – fixing 0.40 hours/m²
(iv) Basic rates
   Plasterboard 1 800 x 1 200 mm @ £7.15/sheet delivered
   Board finish plaster   £78.50/tonne delivered
   Jute scrim in cloth   £4.50/roll (100 m)
   Nails   £1.70/kg
   Plasterboard fixer   £5.53/hour
   Labourer   £4.72/hour

	SYNTHESIS	£	£
(i)	Materials		
	Plasterboard @ £7.15/sheet delivered	7.15	
	Unload 0.10 hours @ £4.72/hour	0.47	
		7.62	
	Nails 0.10 kg @ £1.70/kg	0.17	
	Scrim cloth in 1.80 + 1.20 m		
	= 3 m length per board = $\dfrac{£4.50 \times 3\text{ m}}{100}$	0.14	
		7.93	
	Allow for waste (5%)	0.40	8.33
	1 tonne board finish plaster @ £78.50/tonne delivered	78.50	
	Unload 1 hour @ £4.72/hour	4.72	
		83.22	
	Allow for waste (5%)	4.16	
		87.38 C/F	8.33

B/F   8.33

Allow 25 kg of board finish plaster per 30 m joints;
Therefore cost per board (i.e. 3 m length)

$$= \frac{£87.38 \times 25 \times 3}{1\,000 \times 30}$$

0.22

_____

8.55

Cost of materials for 2.16 m² = £8.55

Therefore cost/m² $= \dfrac{£8.55}{2.16 \text{ m}^2}$ 3.96

(ii)   Labour
Fixing and filling joints of plasterboard
= 0.40 gang hour @ £10.25/hour   4.10

_____

8.06

Net unit rate per m²   **£8.06**

STONE/CONCRETE/QUARRY/CERAMIC TILING/ MOSAIC – SMM7, SECTION M40

Work considered under this section is floor and other finishes that are pre-made solid units of permanent colours in varying range of sizes, shapes and textures, fixed by bonding or adhesion. Some of these materials (e.g. mosaic) are employed to improve appearance and are hard enough to resist abrasion and chemical attack.

ESTIMATING CONSIDERATIONS

Factors affecting cost may include the following.

Type of product

The range of products on the market is large and for wall finish alone there is a selection of ceramics, concrete, terrazzo, slates and stainless steel of various grades, colours, patterns, thicknesses. Likewise concrete, terrazzo and quarry tiles of various sizes, thicknesses and shapes are available for floors; each of these vary in cost and application.

Method of fixing

Methods of fixing the units to their base are many and so are the cost implications. For instance tiles can be bedded in any of the following ways: sand cement, cement-based mortar, mastic adhesive, thick-bed adhesive, all in accordance with the manufacturers' recommendations. Similarly, there is a variety of methods for fixing clay and precast concrete floorings including cement, separating layer, semi-dry, and cement-based adhesive in rubber latex cement mortar and in bitumen emulsion.

Type of work

Patterned work and work requiring a lot of material cutting or fittings such as rounded edge tiles, attached angle tiles, beads and capping slows down labour output and hence costs more to execute.

Location of work

Work may be either external or internal. For example, durable and expensive material such as high-grade ceramics, concrete, glass and stainless steel are specified for external finishes. Moreover an external work is at the mercy of the unpredictable effects of the weather and this factor requires consideration. Similarly labour output is low for work located in awkward positions with accessibility problems.

Nature of background

Weak background has to be made good to ensure proper adhesion. In like manner, prior to fixing the units, floated render should be provided for background which is not plumb and level.

CALCULATING THE NUMBER OF TILES

In building up the unit rate for tile work and slab finishes, it is necessary to determine the number of tiles per square metre. This quantity is calculated using the size of tile and joint thickness specified. However, the number of tiles required can also be calculated by dividing the given area by the area of one tile. The cost of bedding material is included in the unit rate for tile and slab work.

The average labour outputs for tile fixing shown in Table 6.29 are based on a gang of one tradesman and one labourer.

Table 6.29 Guide to average labour output and material coverage for tiling			
Operation	Gang Hours ($/m^2$)	Material coverage	
		Tile (number)	Adhesives (kg)
Fixing quarry tiles			
100 x 100 x 13 mm	1.50	100	–
150 x 150 x 13 mm	1.00	45	–
Fixing terrazzo tile			
225 x 225 x 25 mm	1.00	20	–
300 x 300 x 25 mm	0.90	12	–
Fixing glazed wall tiles			
100 x 100 x 6 mm	0.75	100	1
150 x 150 x 6 mm	0.60	45	1

UNIT RATE BUILD-UP FOR TYPICAL BILL ITEMS

Example 6.45

150 x 150 x 12.5 mm thick clay quarry tile flooring; red; 3 mm joints; symmetrical layout; 10 mm bedding and pointing in cement mortar (1:3); on cement and sand base; to falls only not exceeding 15 degrees from horizontal.   **m²**

DATA
(i)   Basic rates

12 mm thick quarry tile   £140.00/thousand delivered
Portland cement   £56.00/tonne delivered
Sand   £6.00/m³ delivered
Hire of 5/3¹/₂ mixer including fuel and oil   £2.57/hour
Mixer operator   £4.86/hour
Tiler   £5.53/hour
Labourer   £4.72/hour

SYNTHESIS

		£	£
(i)	Materials		
	1 tonne bagged cement @ £56/tonne delivered	56.00	
	Unload 1 hour @ £4.72/hour	4.72	
		60.72	
	1 m³ cement @ £60.72/tonne x 1.442 tonne/m³	87.56	
	3 m³ sand @ £ 6.00/m³	18.00	
		105.56	
	Voids (allow) 25%	26.39	
	Cost of 4 m³ cement and sand mix	131.95	

Therefore cost of 1 m³ = $\dfrac{£131.95}{4}$ 32.99

Cost of mixing 1 m³ mortar
Hire 5/3¹/₂ mixer @ £2.57/hour   = £2.57
Mixer operator @ £4.86/hour   = £4.86

7.43

C/F   40.42

	B/F	40.42	
Allow for compression and waste (10%)		4.04	
Cost of 1 m³ mortar		44.46	

Therefore cost/m² of 10 mm thick bed
= £44.46 x 0.010   0.44

Mortar for jointing and pointing (allow)   0.25   0.69

1 000 tiles @ £140/thousand delivered	140.00	
Unload (allow)	0.75	
	140.75	

Allow for waste (2$^1/_2$%)   3.52

Cost of thousand tiles   144.27

Number of tiles per m² of floor allowing
for 3 mm joints
$$= \frac{1\,000 \times 1\,000}{153 \times 153} = 43 \text{ tiles}$$

Therefore cost of tiles/m²
$$= \frac{43 \times £144.27}{1\,000}$$ 6.20

(ii)   Labour
Laying 10 mm thick bed and tiles
= 1 gang hour @ £10.25/hour   10.25

Net unit rate per m²   **£17.14**

Example 6.46

150 x 150 x 6 mm thick glazed ceramic tiling; coloured; fixing to plastered walls; over 300 mm wide.   **m²**

DATA

(i)   Gang ratio – 1 tiler to 1 labourer

(ii)   Basic rates:

Glazed ceramic tiles   £27.50/100 delivered
Combined tile adhesive and grout   £3.40/kg
Tiler   £5.53/hour
Labourer   £4.72/hour

SYNTHESIS	£	£
(i) Materials		
100 tiles @ £27.50/hundred delivered	27.50	
Unload (allow)	0.20	
Cost of 100 tiles	27.70	
45 number tiles/m² @ £27.70/hundred		
$= \dfrac{45 \times £27.70}{100}$	12.47	
Tile adhesive 1 kg/m² @ £3.40/kg	3.40	
	15.87	
Allow for waste (5%)	0.79	16.66
(ii) Labour		
Fixing tiles 0.60 gang hour @ £10.25/hour		6.15
Net unit rate per m²		**£22.81**

DECORATIVE PAPER/FABRIC – SMM7, SECTION M52

As its name suggests, decorative papers/fabrics are finishes required to decorate surfaces and provide colour or pattern. Like paints, there are many different types (e.g. wallpapers, plastics faced cloths, wood veneers, textiles).

ESTIMATING CONSIDERATIONS

Factors affecting cost may include the following.

Type of decorative paper

There are many different types of paper available each with a different price. Normally vinyl-coated papers, which are washable, are more expensive to purchase and when specified should influence the pricing.

State of background

Where the background to receive the paper is uneven or has some cracks, it requires making good prior to laying the paper. This operation slows down labour output and thus increases the cost of the work.

Plain or pattern work

The amount of labour time spent in matching design in order to bring out the pattern of the work, as well as the waste factor, is more than plain paper work.

Average coverage rates

A standard piece or roll of wall paper is 10.06 x 0.53 m wide, that is, 5.35 m² per roll. Wastage factor is in the region of 10% for plain paper and at least 15% for patterned paper. Allow 1 litre of paste per roll for light paper and 2 litres per roll for heavy and burst paper.

Labour output

Table 6.30 gives the average labour output for wallpapering.

Table 6.30 Guide to average labour output for wallpapering	
Operation per roll	Hours
Sizing of wallpaper	0.25
Hang ordinary paper	1.00
Hang burst paper	1.50

UNIT RATE BUILD-UP FOR TYPICAL BILL ITEM

Example 6.47

Vinyl-coated patterned paper; sizing; applying adhesive; hanging; butt joints; plastered walls and columns; over 0.50 m². m²

DATA
(i) Basic rates:

    Vinyl coated paper   £7.25/roll delivered
    Paste   £1.50/litre
    Paperhanger   £5.53/hour
    Labourer   £4.72/hour

		£	£
	SYNTHESIS		
(i)	Materials		
	1 roll wallpaper @ £7.25/roll delivered	7.25	
	Unload (allow)	0.05	
		7.30	
	1 litre paste per roll @ £1.50/litre	1.50	
		8.80	
	Allow for waste (15%)	1.32	
		10.12	
(ii)	Labour		
	Sizing 1 roll wallpaper 0.25 hour		
	Hanging 1 roll wallpaper 1.00 hour		
	1.25 hours		
	Therefore 1.25 hours @ £5.53/hour	6.91	
		17.03	

Cost of providing and hanging 1 roll of
(5.30 m²) wallpaper = £17.03

Therefore cost/m² = $\dfrac{£17.03}{5.30}$ 3.21

Net unit rate per m²   **£3.21**

PAINTING/CLEAR FINISHING – SMM7, SECTION M60

Painting is the application to surfaces of pigmented liquids or semi-liquids which, on drying, protect components from rain, sunlight, abrasion, chemical liquids and fumes, corrosion, fungi, insects and fire. Materials used in painting are many but an ordinary paint system comprises a primer, undercoat and finishing coats which provide the final protection against the adversaries mentioned above. With its protective capacity, paintwork is required on new work as well as existing, both indoors and outdoors.

ESTIMATING CONSIDERATIONS

Factors influencing cost of paintwork may include the following:

Nature of background

The covering capacity of paint, varnish and similar materials depends on the porosity and texture of the surface to which it is applied. Concrete and brick absorb more paint than plastered surfaces. At the same time, the nature of the background would reflect the brush drag and affect productivity.

Complexity of work

Paintwork in multicolours, on surfaces of intricate finish in multicolours, to glazed surfaces in small pane areas, or to work in narrow widths, is time-consuming and hence escalates costs.

Location of work

Paintwork in confined spaces, to exposed locations high above ground, and to locations necessitating working from ladders, reduces labour output and pushes up the cost. Work on very high ceilings may require expensive temporary work such as a mobile scaffold tower or birdcage scaffold.

Preparation required

The amount of preparation required prior to painting (e.g. masking, washing and rubbing down, burning off and sandpapering, wire brushing and dusting off, knotting, priming and stopping between coats, rubbing down and removing distempers) reduces output and hence increases costs.

Method adopted

In some situations, applying the paint by spraying can be faster and cheaper than brush application; however, spraying requires more preparatory work such as masking and cleaning.

Number of coats

The number of coats applied to a surface determines the quantity of materials and labour time required.

Type of paint specified

A wide range of paints is available, each type performing a specific protective function. Their variety demonstrates their price range, e.g. paint chosen to perform a fire retarding function will be more expensive on the market than paint required to provide hygienic, aesthetic or colour improvement functions.

Pricing

For reasons of convenience, the price synthesis for paintwork is prepared by using large quantities, and reducing to individual unit rates on completion.

Labour output

Tables 6.31 and 6.32 are guides to the average labour output for painting work.

Average coverage rates

Table 6.33 gives average coverage rates for various painting materials.

Table 6.31 Guide to average labour output for painting general surfaces of walls and ceilings

Operation	Painter hours per 100 m²				
	Brick	Concrete		Plaster	
	Walls	Walls	Ceilings	Walls	Ceilings
Emulsion paint	10	10	13	8	10
Oil-bound primer	14	13	17	10	14
Oil-based undercoat	17	13	20	11	17
Oil-based finishing coat	20	14	25	13	20
Cement based paint	14	13	–	10	–
Corrugated surfaces	Add 12$^1/_2$%				
Narrow widths	Add 50%				
Multi-coloured works	2 Colours – add 10%				
	3 Colours – add 15%				

Table 6.32 Guide to average labour output for painting metal surfaces

Operation	Painter hours per 100 m²
General metal surfaces	
Wire brushing	14
Prepare and prime	20
Touch-up primer and 1 undercoat	17
Finishing coat	16
Mordant solution	20
Bituminous paint	13
Structural steel	Above outputs plus 12$^1/_2$%
Roof trusses/radiators	Above outputs plus 50%

Table 6.33 Covering capacities of paint

Litres required for 100 m²

Operation	On brick surfaces	On plaster surfaces	On wood surfaces	On metal surfaces	On concrete surfaces
Washable distemper	(10 kg) 8$\frac{1}{2}$	(15 kg)			
Emulsion paint		8			8
Oil-based primer	14	10	10	10	13
Oil-based undercoat	10	8$\frac{1}{2}$	8$\frac{1}{2}$	8$\frac{1}{2}$	9
Oil-based finishing coats	5–8	5–8	5–8	5–8	5–8
Varnish			8		
Wood Stain					
(oil-based)			7		
(alcohol-based)			3$\frac{1}{2}$		
Knotting			3–4		
Stopping			2$\frac{1}{2}$ kg putty		
Glass paper			28 sheets	For 4 coats of paint = 7 number sheets per coat per 100 m²	

UNIT RATE BUILD-UP FOR TYPICAL BILL ITEMS

Example 6.48

> **Preparing; applying; three coats of vinyl emulsion paint; concrete general surfaces; over 300 mm girth.**   **m^2**

DATA
(i)   Basic rates:
     Emulsion paint   £8.50/5 litres delivered
     Painter   £5.53/hour

SYNTHESIS

(Based on 100 m^2 surface area)

	£	£
(i) Materials		
Emulsion paint @ £8.50/5 litres delivered	8.50	
Unload (allow)	0.10	
	8.60	
Allow for waste (5%)	0.43	
Cost of 5 litres of paint	9.03	

Cost of three coats of paint for 100 m^2:

First coat	8 litres
Second coat	8 litres
Third coat	8 litres
	24 litres @ £9.03/5 litres

$$= \frac{24 \times £9.03}{5} \qquad 43.34$$

(ii)   Labour
     Applying 3 coats of emulsion paint over 100 m^2
       surface area at 10 hours/coat = 30 hours
       @ £5.53/hour   165.90

Material and labour cost/100 m^2 surface area   209.24

Therefore cost of 1 m$^2 = \dfrac{£209.24}{100 \text{ m}^2}$ 2.09

Net unit rate per m^2   **£2.09**

Example 6.49

Preparing; knotting; one coat primer; two undercoats and one coat finishing gloss on softwood door; over 300 mm girth.   m²

DATA
(i) Basic rates:

Glasspaper   £0.14/sheet
Putty   £0.50p/kg
Knotting   £4.75/litre delivered
Wood primer   £13.00/5 litres delivered
Undercoat   £12.00/5 litres delivered
Gloss   £12.00/5 litres delivered
Painter   £5.53/hour

SYNTHESIS

(Based on 100 m² surface area)

(i) Materials

	£	£
7 Sheets glass paper @ £0.14/sheet	0.98	
2.5 kg putty @ £0.50/kg	1.25	
0.75 litre knotting @ £4.75/litres delivered	3.56	
10 litres wood primer @ £13.00/5 litres delivered	26.00	
2 x 8.5 litres undercoat @ £12.00/5 litres delivered	40.80	
8 litres gloss @ £12.00/5 litres delivered	19.20	
	91.79	
Allow for unloading and waste (6%)	5.51	
	97.30	

(ii) Labour

Prepare and prime	=	20 hours	
Undercoat 2 x 17	=	34 hours	
1 finishing coat	=	16 hours	
		70 hours @ £5.53/hour	387.10

Material and labour cost/100 m² surface area   484.40

Therefore cost of 1 m² = $\frac{£484.40}{100}$ 4.84

Net unit rate per m²   **£4.84**

PLUMBING INSTALLATION: DISPOSAL AND PIPED SUPPLY SYSTEMS - SMM7, SECTIONS R10 AND S10

Work considered under this section is as follows:

1. Disposal systems (SMM7, Section R10)
2. Piped supply systems (SMM7, Section S10)

Work required in these two sections is executed by a plumber assisted by a mate.

Disposal systems

Disposal systems comprise rainwater gutters and general pipework, fixtures, and so forth, used in rainwater and foul drainage above ground.

MATERIALS USED

Materials commonly used for this type of work are as follows.

Cast iron

In the past, cast iron was the usual material employed for this type of work. Whilst cast iron is a robust material, it is heavy, prone to fracture and has high initial and maintenance costs.

Plastics

In recent years, plastics have rapidly replaced cast iron in popularity because pipes and gutters made of plastics are light in weight, easy to fix and require no painting.

Aluminium

1. *Cast aluminium*   Cast aluminium is strong, rustless, light in weight and can be purchased in longer lengths.
2. *Extruded aluminium*   This is used for gutter work.

Galvanized mild steel

This can be used in both gutter and pipework. It is light in weight but subject to corrosion if the galvanizing is damaged.

Gutters and pipes in the above material can be obtained in the shapes and lengths listed in Table 6.34.

	Table 6.34 Types and shapes of gutters and pipes			
Item	Cast iron	Plastic	Aluminium	Galvanized steel
Gutter	Shape			
	Half-round ogee Box Moulded Valley	Half-round Square	Half-round Ogee Box Moulded Bevelled	Box Valley
	Length			
	Up to 2.0 m	Up to 4.0 m	Up to 3.0 m	Up to 3.0 m
Pipe	Shape			
	Round Square Rectangular	As cast iron	As cast iron	As cast iron
	Length			
	Up to 2.0 m	Up to 5.5 m	Up to 6.0 m	Up to 3.0 m

ESTIMATING CONSIDERATIONS

Factors that may influence the cost of disposal systems are as follows.

Weight of material

Cast iron is a heavy, brittle material and is thus difficult to work with in terms of unloading, transporting and lifting into position. The estimator will therefore have to make sufficient allowance in his or her price to cover these difficulties.

Presence of obstructions

Obstructions in the vicinity of the work area (e.g. irregular-shaped roof, projections on walls) increase the number of joints in the work. Joint making involves cutting away sections of material; especially in cast iron gutter work, some drilling, extra bolts and nuts are required and hence additional time and material are expended on the work.

Location of work

Work located in a built-up area or at a great height from ground level will require extra safety measures to safeguard the operations and the general public. Hoisting time will need to be considered for work located in these situations.

Ease of accessibility

Work located in an awkward place with an accessibility problem for materials, tools and operatives will hinder productivity and the estimator should make an allowance in his or her price for this problem.

Cast iron rainwater pipe

Projecting ear or earless cast iron rainwater pipe may be specified in the bill of quantities. In the case of earless pipes, the estimator should allow one number holderbat for each length of pipe. He or she should also note whether pipes with projecting ears are intended for fixing direct to wall. If not an allowance should be made in the price for distance pieces.

Allowance for displacement by fittings

Fittings are measured as 'extra over' in the bill of quantities. In practice deductions are made for gutter/pipe displaced by fittings as shown in Table 6.35.

Extras to gutter fitting

A fitting to a cast iron gutter requires one extra fixing bracket, screws, gutter bolt and nut, and a jointing mastic.

Table 6.35 Allowance for displacement by fittings	
Product	Allowance (m)
Gutters	
Angles	0.45
Outlet	0.30
Pipes	
Bends	0.30
Shoes	0.15
Single branch	0.45
Single junctions	0.45
·Double junctions	0.60
Swanneck bends	
225 mm projection	0.60
300 mm projection	0.65

Extras to pipe fitting

A pipe fitting requires one extra fixing and joint; however, junctions require two extra joints.

Gang working

A plumber always works with a mate who is usually an apprentice and paid a labourer's rate. A joint rate for plumber and mate is used for calculating the unit rates.

Table 6.36, as a guide to labour output, is therefore based on gang hours of plumber and mate per linear metre of 2-metre lengths of gutter fixed to timber or pipe fixed to masonry.

Piped supply systems

Work items considered under this section are pipework for cold water installation, which covers pipework from supply authority's stopcock to water utilization points. This involves some underground work (i.e. trench excavation), under-floor work and, at times, work in ducts and in roof space.

Table 6.36 Guide to average labour output for disposal systems work				
	Per linear metre	Fittings (each)		
Gutters		Angle	Outlet	Stop-end
Cast iron				
75 mm half-round	0.30	0.30	0.30	0.30
100 mm half-round	0.50	0.50	0.50	0.50
150 mm half-round	0.75	0.75	0.75	0.75
Aluminium				
75 mm half-round	0.25	0.25	0.25	0.25
100 mm half-round	0.40	0.40	0.40	0.40
150 mm half-round	0.60	0.60	0.60	0.60
Plastic				
75 mm half-round	0.20	0.20	0.20	0.20
100 mm half-round	0.25	0.25	0.25	0.25
150 mm half-round	0.40	0.40	0.40	0.40
Pipes		Bend	Shoe	Single branch
Cast Iron				
75 mm diameter	0.60	0.60	0.60	0.60
100 mm diameter	0.75	0.75	0.75	0.75
150 mm diameter	1.00	1.00	1.00	1.00
Aluminium				
75 mm diameter	0.50	0.50	0.50	0.50
100 mm diameter	0.75	0.75	0.75	0.75
150 mm diameter	0.95	0.95	0.95	0.95
Plastic				
75 mm diameter	0.40	0.35	0.35	0.35
100 mm diameter	0.50	0.45	0.45	0.45
150 mm diameter	0.75	0.55	0.55	0.55

MATERIALS FOR PIPEWORK

Materials used in this type of work are copper, polythene or plastic and steel tubes and fittings.

MATERIALS FOR PIPEWORK

Materials used in this type of work are copper, polythene or plastic and steel tubes and fittings.

Copper tubes

These, in general, are expensive materials, have a high resistance to corrosive agents, and are of three types:

1. *Soft underground copper* This is for use in underground mains and is obtained in coils up to 20 m long.
2. *Light gauge* This is for use in cold and hot water installations and is obtained in 6 m lengths.
3. *Thin wall* This type should not be bent or buried. Elbows should be used whenever it is employed. It is obtained in 6 m lengths.

Polythene tube

This is supplied in various standard coils up to 165 m long.

Plastic tube

This is used in wastes and overflows.

Steel pipes

Steel pipes are cheaper, stronger and require less support for installation. They are used in cold and hot water pipework and may be categorized as follows:

1. *Galvanized mild steel*
2. *Stainless steel* This is more resistant to corrosion than copper pipes.

Fittings

All fittings to the above tubes are measured as 'extra over' in the bill of quantities. Joints to fittings are made as follows:

1. *Copper to fittings* By compression, capillary, welding or brazing joint.
2. *Polythene to fitting* By compression with an 'O' ring.
3. *Steel to fitting* By screwing and socketing.
4. *Plastic to fitting* By connectors with solvent.

ESTIMATING CONSIDERATIONS

Factors affecting cost of cold water installations are as follows.

Accessibility

Lengths of pipes used in internal work are determined by height of walls, ceilings and the accessibility to under-floor and roof spaces. A lack of accessibility or inadequate headroom will cause a lot of abortive work. For instance, prepared pipes will need several adjustments, and even a fixed pipe may have to be removed for adjustment for final fixing. The estimator will have to consider this and make an allowance for it in the price.

Number of joints

The length of pipe that can be installed at any one time will depend upon:

1. Use of pipe.
2. Lengths of pipe available.
3. Ease of handling.
4. Accessibility to work areas.

Where problems cause many joints to be made in the running length, this consumes time and material, and the estimator will have to bear this in mind when he or she is considering the number of joints to allow in the price for the installation.

Work in short lengths

This work is time-consuming, and hence the price for the operation output should be increased accordingly.

Obstructions

Work that is more awkward by reason of obstructions and which calls for more bends, tees and elbows in the running length of pipe, and so on, slows down the tempo of work and an allowance should be made to cover these extra jointing operations.

Labour output

Table 6.37 gives a guide to combined hours for plumber and mate on the installation of piped supply systems.

Table 6.37 Guide to average labour output for piped supply systems work					
Nominal diameter (mm)	To walls	To trench	Fittings		
			Made bend	Elbows	Tee
Copper tubing					
15	0.25	0.10	0.20	0.20	0.35
22	0.25	0.15	0.30	0.25	0.40
28	0.30	0.20	0.40	0.30	0.45
35	0.35	0.25	0.50	0.35	0.50
44	0.40	0.30	0.60	0.40	0.55
Polythene tubing					
13	0.08	0.06	0.08	0.08	0.12
19	0.12	0.08	0.12	0.12	0.16
25	0.15	0.10	0.16	0.16	0.20
32	0.18	0.12	0.20	0.20	0.25
40	0.20	0.14	0.25	0.25	0.30

UNIT RATE BUILD-UP FOR TYPICAL BILL ITEMS

Example 6.50

> Cast iron rainwater pipe and fittings; 100 mm; jointed in cold joint compound; ears fixing to masonry with galvanized pipe nails and distance pieces. **m**

DATA
(i) Basic rates:
100 mm diameter cast iron pipe   £39.50/2 m length delivered
Fill plug   £11.25/kg
Plumber   £5.53/hour
Plumber's mate   £4.72/hour

SYNTHESIS

	£	£
(i) **Materials**		
100 mm diameter cast iron pipe @ £39.50/2 m delivered	39.50	
Unload 0.05 hour @ £4.72/hour	0.24	
	39.74	
0.20 kg fill plug @ £11.25/kg	2.25	
2 number distance pieces (allow)	0.30	
2 number galvanized pipe nails (allow)	0.30	
	42.59	
Allow for waste (5%)	2.13	
Cost of 2 m length pipe	44.72	
Therefore cost/m $= \dfrac{£44.72}{2}$		22.36
(ii) **Labour**		
Plumber £5.53		
Mate £4.72		
£10.25		
0.75 hour (plumber and mate) @ £10.25/hour	7.69	
		30.05

Net unit rate per m   **£30.05**

Example 6.51

Extra over 100 mm diameter cast iron pipe for bend.   Nr

DATA
(i) Basic rates:
   100 mm diameter cast iron pipe   £39.74/2 m length delivered
   100 mm diameter cast iron bend   £9.25 each delivered
   Fill plug   £11.25/kg
   Plumber   £ 5.53/hour
   Plumber's mate   £4.72/hour

SYNTHESIS
		£	£
(i)	Materials		
	1 number 100 mm diameter cast iron bend @ £9.25 each delivered		9.25
	Unload (allow)		0.05
			9.30
	Less displacement of 300 mm length of pipe $= \dfrac{£39.74}{1} \times \dfrac{300 \text{ mm}}{2\,000 \text{ mm}}$		5.96
			3.34
	0.20 kg fill plug @ £11.25/kg		2.25
	1 number distance piece (allow)		0.15
	2 number galvanized pipe nails (allow)		0.30
			6.04
	Allow for waste (5%)		0.30
			6.34
(ii)	Labour		
	0.75 hour (plumber and mate) @ £10.25/hour		7.69
			14.03

Net unit rate per Nr   **£14.03**

Example 6.52

> UPVC rainwater pipe; 65 mm diameter; jointed with socket connectors; fixing to masonry with brackets; plastic distance pieces and galvanized screws. **m**

DATA
(i) Basic rates:

      65 mm diameter upvc bend   £8.75/4 m length delivered
      Brackets   £0.66 each
      Socket connectors   £0.81 each
      Plumber   £5.53/hour
      Plumber's mate   £4.72/hour

SYNTHESIS

		£	£
(i)	Materials		
	65 mm diameter upvc pipe @ £8.75/4 m length delivered	8.75	
	Unload (allow)	0.15	
		8.90	
	2 number plastic distance pieces (allow)	0.30	
	1 number socket connector @ £0.81 each	0.81	
	1 number fixing bracket @ £0.66 each	0.66	
	Galvanized screws and plugs (allow)	0.05	
		10.72	
	Allow for waste (5%)	0.54	
	Cost of 4 m length pipe	11.26	
	Therefore cost/m = $\dfrac{£11.26}{4}$	2.82	
(ii)	Labour		
	0.40 hour (plumber and mate) @ £10.25/hour	4.10	
		6.92	

Net unit rate per m   **£6.92**

Example 6.53

Cast iron eaves gutter and fittings; 100 mm half-round; bolted and jointed in cold joint compound; securing to softwood with fascia brackets and galvanized screws.   **m**

DATA

(i)   Basic rates:

100 mm half-round cast iron gutter   £15.75/2 m length delivered
Fill plug   £11.25/kg
Fascia brackets   £0.48 each
Plumber   £5.53/hour
Plumber's mate   £4.72/hour

		£	£
SYNTHESIS			
(i)	Materials		
	100 mm half-round cast iron gutter @ £15.75/2 m		
	length delivered	15.75	
	Unload (allow)	0.15	
		15.90	
	2 number fascia brackets @ £0.48 each	0.96	
	0.08 kg fill plug @ £11.25/kg	0.90	
	Bolts and screws (allow)	0.06	
		17.82	
	Allow for waste (5%)	0.89	
	Cost of 2 m length gutter	18.71	
	Therefore cost/m = $\dfrac{£18.71}{2}$		9.36
(ii)	Labour		
	0.75 hour (plumber and mate) @ £10.25/hour		7.69
			17.05

Net unit rate per m   **£17.05**

Example 6.54

Extra over 100 mm half-round cast iron gutter for 100 mm running outlet. **Nr**

DATA
(i)   Basic rates:
      100 mm cast iron outlet   £4.50 each delivered
      100 mm half-round cast iron gutter   £15.75/2 m length delivered
      Fill plug   £11.25/kg
      Fascia brackets   £1.25 each
      Plumber   £5.53/hour
      Plumber's mate   £4.72/hour

	SYNTHESIS	£	£
(i)	Materials		
	1 number 100 mm cast iron outlet @ £4.50 each		
	delivered	4.50	
	Unload (allow)	0.05	
		———	
		4.55	
	Less displacement of 100 mm half-round gutter		
	$\dfrac{£15.75}{1} \times \dfrac{300\text{ mm}}{2\,000\text{ mm}} =$	2.36	
		———	
		2.19	
	1 number fascia bracket @ £1.25 each	1.25	
	0.20 kg fill plug @ £11.25/kg	2.25	
	Bolts and screws (allow)	0.06	
		———	
		5.75	
	Allow for waste (5%)	0.29	
		———	
			6.04
(ii)	Labour		
	0.50 hour (plumber and mate) @ £10.25/hour	5.13	
			———
			11.17
			═══

Net unit rate per Nr **£11.17**

Example 6.55

Copper pipes; 22 mm diameter; capillary fittings; fixed to masonry with clips.   m

DATA
(i) Basic rates:
    22 mm diameter copper tube   £18.00/6 m length delivered
    Copper saddle clips   £12.00/100 number
    22 mm straight couplings   £0.90 each (allow $^1/_3$ m length
      of pipe)
    Plumber   £5.53/hour
    Plumber's mate   £4.72/hour

SYNTHESIS

	£	£
(i) Materials		
22 mm diameter copper tube @ £18/6 m length delivered	18.00	
6 number saddle clips @ £12.00/100 number	0.72	
Screws and plugs (allow)	0.50	
Straight coupling £0.90 x 6 x 33$^1/_3$ %	1.80	
Jointing material (allow)	0.25	
	21.27	
Allow for waste (2$^1/_2$%)	0.53	
Cost of 6 m length pipe	21.80	
Therefore cost/m = $\dfrac{£21.80}{6}$	3.63	
(ii) Labour		
0.25 hour (plumber and mate) @ £10.25/hour	2.56	
	6.19	

Net unit rate per m   **£6.19**

Example 6.56

| Extra over 22 mm diameter copper pipe for equal tee. | Nr |

DATA
(i) Basic rates:
   22 mm diameter equal tee   £1.80 each delivered
   22 mm diameter copper tubing   £3.00/m delivered
   Plumber   £5.53/hour
   Plumber's mate   £4.72/hour

SYNTHESIS	£	£
(i) Materials		
1 number 22 mm diameter equal tee @ £1.80		
each delivered	1.80	
Less 75 mm of tubing @ £3.00/m		
$= \dfrac{£3.00}{1} \times \dfrac{75\text{ mm}}{1\,000\text{ mm}}$	0.23	
	2.03	
Allow for waste (2^1/$_2$%)	0.05	
	2.08	
(ii) Labour		
0.40 hour (plumber and mate) @ £10.25/hour	4.10	
	6.18	

Net unit rate per Nr   **£6.18**

Example 6.57

Polythene tube; 25 mm diameter; in trench; straight.	m

DATA
(i) Basic rates:
25 mm diameter polythene tube £21.00/30 m
length delivered
Plumber £5.53/hour
Plumber's mate £4.72/hour

SYNTHESIS	£	£
(i) Materials		
25 mm diameter polythene tube @ £21/30 m delivered	21.00	
Straight connector (allow) 1 number @ £0.95 each	0.95	
	21.95	
Allow for waste (2½%)	0.55	
Cost of 30 m length pipe	22.50	
Therefore cost/m = $\dfrac{£22.50}{30}$	0.75	
(ii) Labour		
0.10 hour (plumber and mate) @ £10.25/hour	1.03	
	1.78	

Net unit rate per m	**£1.78**

Example 6.58

Extra over 25 mm polythene tubing for brass compression type elbow. **Nr**

DATA
(i) Basic rates:
    25 mm diameter elbow   £3.70 each
    Plumber   £5.53/hour
    Plumber's mate   £4.72/hour

SYNTHESIS	£	£
(i) Materials		
1 number 25 mm diameter elbow @ £3.70 each	3.70	
	3.70	
Allow for waste (5%)	0.19	
	3.89	
(ii) Labour		
0.16 hour (plumber and mate) @ £10.25/hour	1.64	
	5.53	

Net unit rate per Nr   **£5.53**

DRAINAGE BELOW GROUND – SMM7, SECTION R12

Work considered under this section comprises drain trenches, concrete bed, haunching and surrounds, pipes and accessories, manholes, inspection chambers, soakaways, cesspits and septic tanks.

DRAIN TRENCHES

The unit of measurement is the linear metre. However, the rate for excavating drain trenches includes excavation, earthwork support, disposal of excavated material, filling to excavations, and compacting bottoms of excavations. The principles applied in unit rate build-up for drain trenches are the same as the unit rate build-up for the various items of work treated in the groundwork section (p. 102).

The estimator is required to sketch a typical section of trench and take quantities for pricing. The depth of trench is normally given. However the width of trench is unknown and is at the estimator's discretion. The usual minimum widths of trenches for pipes up to 200 mm diameter are shown in Table 6.38.

Table 6.38 Minimum widths of trenches

Depth up to (m)	Minimum width (mm)
1.00	500
1.50	700
3.00	800
4.50	900

The minimum width of trench for pipes over 200 mm diameter would increase by an average of an additional diameter of the pipe. However, for machine excavation, the minimum trench width is the width of the machine bucket.

The average labour output for surface trench excavation and factors influencing cost considered previously (pp. 102–9) applies equally to drain trench excavation.

CONCRETE BEDS, HAUNCHING AND SURROUNDS

The unit of measurement for concrete beds, haunching and surrounds is the linear metre with the mix, sectional area of concrete, and diameter of pipe stated. The principles involved in unit rate build-up of concrete work under this section is basically the same as the concrete work already discussed, that is, synthesis of material, mixing and placing costs (see pp. 118–25). The unit rate build-up for linear metre of concrete beds and so forth will involve the calculation of the cost per cubic metre of placed concrete; this is then converted to cost per linear metre and an allowance made for labour when working concrete in small quantities in a trench.

In practice, an average material coverage will be given in tables (e.g. concrete bed, benching surround). However, the formula for determining the width of concrete bed is as follows:

Width of bed = external diameter of pipe + 300 mm

For example:

Width of bed for 100 mm diameter pipe = 100 + 25 (twice wall thickness) + 300 mm = 425 mm.

Average material coverage and displacement of backfill are shown in Tables 6.39 and 6.40.

Table 6.39 Average amount of concrete in drain beds, haunching and surrounds

	Pipe diameter (mm)			
	100	150	225	300
Size of concrete bed	425 x 100	475 x 100	550 x 150	625 x 150
	Concrete per metre run (m³)			
Bed only	0.043	0.043	0.083	0.095
Bed and haunch	0.058	0.066	0.103	0.114
Bed and 150 mm surround	0.150	0.181	0.103	0.312

Note: Volume of concrete bed displaces same volume of backfill.

Table 6.40	Displacement of backfill by volume of pipe per metre run			
	Pipe diameter (mm)			
	100	150	225	300
Per metre run (m³)	0.013	0.025	0.052	0.085

PIPES

The materials incorporated in the manufacture of pipes used for drainage below ground may be concrete, iron, steel, vitrified clay, upvc, pitch fibre, and so on, depending on the design of drainage which may be either rigid or flexible. The unit of measurement is linear and measured over all joints. However, fittings are enumerated and measured extra over pipe lengths.

ESTIMATING CONSIDERATIONS

Factors affecting cost of pipework below ground may be summarized as follows.

Depth of pipe trenches

Laying pipes in very deep trenches slows down labour output. Operatives work in confined spaces in deep trenches with problems of accessibility for operatives and materials.

Complexity of drainage layout

Complex drainage systems incorporating several short branches requires a substantial amount of cutting and hence an increase in the wastage factor. At the same time, the ratio of jointing coupling to pipe increases as the pipe joints increase (as a result of jointing shorter pipe lengths), leading to a corresponding rise in cost of the pipework.

Flexibility of design

Where extra flexibility and deflection are required, pipe joints are increased with the insertion of expansion-joint support to achieve this aim. This increases the cost of pipes laid for the reasons given above.

Location of pipework

Drain pipes may be located inside a building and within or without the boundary of a site. Each location may call for a different pipe material. For example, drainage under main roads may be in cast iron, either encased in concrete or in deeper trenches, thus increasing the cost of material and labour. Suspended overhead and vertical work reduces operatives' output as well as increasing the cost by the provision of supports for such work.

Market prices for pipes

Prices of pipes are normally quoted plus or minus a percentage on or off a basic price list depending on the quantities bought. It therefore follows that better purchasing terms can be obtained in bulk purchases. The quality of vitrified clayware pipe reflects its price as follows:

Best quality	– basic price list
British Standard	– basic price list plus 10 per cent
British Standard tested	– basic price list plus 50 per cent
Second quality	– basic price list less 10 per cent

Pipes can be purchased in the lengths and sizes shown in Table 6.41.

Labour output

Table 6.41 gives a guide to average labour output on pipe laying and jointing.

Table 6.41   Pipes – availability in lengths and sizes

Pipe type	Length (m)	Diameter (mm)
Vitrified clay	1.6	100–400
Pitch fibre	1.5, 2.5, 3.0	75–100
UPVC	3.0, 6.0	110–160
Concrete	2.0	180–600
Spun iron	2.0, 3.0, 4.0	75–225

Table 6.42   Guide to average labour output for below-ground drainage work

Pipe laying and jointing	Diameter of pipe (mm)				
	75	100	150	225	300
Vitrified clay pipe					
Laying per m		0.07	0.10	0.15	0.18
Laying per 3 m length		0.10	0.12	0.17	0.20
Laying bends or tapers each		0.06	0.08	0.11	0.13
Jointing each		0.05	0.08	0.11	0.17
Pitch fibre pipe					
Laying per m	0.03	0.04	0.06	0.08	
Laying bends or tapers each	0.06	0.07	0.08	0.10	
Jointing each	0.04	0.05	0.08	0.10	
Cutting each	0.03	0.04	0.06	0.08	
Cast iron pipe					
Laying per m		0.55	0.72	1.10	1.70
Laying per 3 m length		0.72	1.10	1.60	2.00
Laying bends or tapers each		0.50	0.67	1.00	1.50
Jointing each		0.25	0.50	1.00	1.50
UPVC pipe					
Laying per 3 m length	–	0.15	0.20	–	–
Jointing each	–	0.10	0.15	–	–

Joints and pipe displacement

Table 6.43 gives the number of joints created by a pipe fitting and the length of pipe displacement.

Table 6.43 Displacement and joints in fittings

Type of pipe	Fitting			
	Bend	Taper	Single junction	Double junction
Vitrified clay pipe				
Number of joints	1	1	1–2	2–3
Displacement (m)	0.50	0.30–0.60	1.0	1.20
Pitch fibre pipe				
Number of joints	1	–	2	3
Displacement (m)	1.22	–	0.60	1.00
Cast iron pipe				
Number of joints	1	1	1–2	2–3
Displacement (m)	0.46	0.39	0.53	0.69

MANHOLES AND SEPTIC TANKS

Work in connection with provision of manholes and septic tanks involves excavation, concrete work, masonry and other sundry items. These sections have been previously described and factors influencing the cost of work in excavation, concrete work and masonry apply equally to provision of manholes. However, it is prudent to increase the average labour output by about 10–25 per cent to allow for restricted working space below ground.

UNIT RATE BUILD-UP FOR TYPICAL BILL ITEMS

Example 6.59

> Excavating trenches to receive pipes not exceeding 200 mm girth; commencing from natural ground level; disposing of surplus excavated material off site; average depth 1.50 m.   m

DATA
(i)   Trench in ordinary ground
(ii) Size of trench 1.00 m length x 0.70 m width x 1.50 m deep
(iii) Machine excavation
(iv) 75% of excavated material backfilled
(v)   Average output:
      Labour forming gradients to bottom 6 m^2/hour
      Labour filling to excavation 1 m^3/hour
      Machine excavates 5 m^3/hour
(vi) Basic rates:
      Hire of excavator (including driver, fuel, etc.)
         £15.00/hour
      Labourer   £4.72/hour
      Banksman   £4.72/hour

SYNTHESIS   £   £
(i)   Excavation
      Quantity of excavation per 1 metre of trench
         1.00
         0.70
         1.50
         ───
               1.05 m^3

      Hourly cost of excavator and banksman
         = £15.00 + £4.72 = £19.72
      Cost of excavating 1 metre of trench
         $= \dfrac{£19.72 \times 1.05\ m^3}{5\ m^3}$ 4.14

(ii) Earthwork support
      Amount required per 1 metre of trench
         2/1.00
           1.50
           ───
               3.00 m^2 (allow) @ £1.80/m^2 5.40
                                                        ──────
                                              C/F   9.54

| | | B/F | 9.54 |

(iii) Forming bottoms to gradients
Area prepared per 1 metre of trench

1.00
0.70
———

$$0.70 \text{ m}^2 = \frac{\pounds4.72 \times 0.70 \text{ m}^2}{6 \text{ m}^2/\text{hour}}$$ 0.55

(iv) Filling
75% excavated material backfilled
= 1.05 m³ x 75% = 0.79 m³ @ £4.72/m³   3.73

(v) Compact filling
Area 0.70 m² @ £0.40 per m²   0.28

(vi) Disposal of excavated material off site
Volume of excavated material disposed of
= 25% of 1.05 m³ x 1.25 (for 25% bulkage)
= 0.33 m³

Therefore 0.33 m³ @ £4.66/m³ (as Example 6.7)   1.54
_____
15.64
======

Net unit rate per m   **£15.64**

Example 6.60

Plain *in situ* concrete (1:2:4–20 mm aggregates) bed 475 x 100 mm; poured on or against face of earth.   m

DATA
(i)   Net price of concrete in foundations (as Example 6.9)
           = £52.45/m^3
(ii) Basic rates:
       Labourer £4.72/hour

SYNTHESIS   £   £
(i)   Materials
      Volume of concrete per 1 metre of trench
           1.00
           0.48
           0.10
           ─────
                 0.048 m^3 @ £52.45/m^3 2.52

(ii) Labour
      Allow for extra labour working concrete in small
          quantities in drain trench 0.50 hour/m @
          £4.72/hour   2.36
                                                          ─────
                                                           4.88
                                                          ═════

Net unit rate per m   **£4.88**
                                                                    ═════

Example 6.61

Drains; 150 mm diameter vitrified clay pipes and fittings BS.65; pipework in trenches; cement mortar (1:2) and tarred gaskin joints.   **m**

DATA
(i)   Vitrified clay pipe taken as British Standard quality
(ii) 5/3^1/$_2$ mixer output = 1 m^3/hour
(iii) Gang ratio – 1 pipe layer to 1 labourer
(iv) Basic rates:

Portland cement   £56.00/tonne delivered
Sand   £6.00/m^3 delivered
150 mm diameter clay pipe   £8.75 per piece (1.60 m) delivered
Hire 5/3^1/$_2$ mixer (including fuel and oil)   £2.57/hour
Pipelayer   £4.86/hour
Mixer operator   £4.86/hour
Labourer   £4.72/hour

SYNTHESIS

	£	£
(i) Materials		
1 tonne bagged cement @ £56/tonne delivered	56.00	
Unload 1 hour @ £4.72/hour	4.72	
	60.72	
1 m^3 cement @ £60.72/tonne x 1.442 tonne/m^3	87.56	
2 m^3 sand @ £ 6.00/m^3	12.00	
	99.56	
Voids (allow) 25%	24.89	
Cost of 3 m^3 cement and sand mortar	124.45	
Therefore cost/m^3 = $\dfrac{£124.45}{3}$	41.48	
Cost of mixing 1 m^3 of mortar		
Hire 5/3^1/$_2$ mixer 1 hour @ £ 2.57/hour = £2.57		
Mixer operator 1 hour @ £4.86/hour = £4.86		
	7.43	
	C/F	48.91

		B/F	48.91
	Allow for waste (5%)		2.45
			———
	Cost of 1 m³ mortar		51.36
			═══

(ii) Clay pipe
1.6 m length @ £8.75 each delivered 8.75
Less 5% (quantity) 0.44

——
8.31
Unload (allow) 0.20

——
8.51
Allow for waste (2^1/2%) 0.21

——
8.72
Gaskin (allow) 0.40

——
9.12

Mortar joint
0.0015 m³ cement mortar/joint @ £51.36/m³ 0.08

——
9.20

(iii) Labour
Laying   0.12 gang hour
Jointing   0.08 gang hour
              ————
              0.20 gang hour @ £9.58/hour 1.92

——
Cost of laying 1.60 m length pipe 11.12
═══

Therefore cost/m $= \dfrac{£11.12}{1.60 \text{ m}}$ 6.95

——

Net unit rate per m   **£6.95**
═══

Example 6.62

Extra; bends   **Nr**

DATA
(i) Gang ratio – 1 pipelayer to 1 labourer
(ii) Basic rates:
     150 mm bend   £4.75 each delivered
     150 mm diameter clay pipe   £8.75 per piece
       (1.60 m) delivered
     Pipelayer   £4.86/hour
     Labourer   £4.72/hour

SYNTHESIS   £   £

(i) Materials
     1 number 150 mm bend @ £4.75 each delivered   4.75
     Unload (allow)   0.10
                                                    ─────
                                                    4.85

     Adjustment for bend being extra over price
     Deduct 0.50 metre 150 mm pipe @
       (£8.75 less 5% = £8.31 ÷ 1.6 m) £5.19/m   2.60
                                                    ─────
                                                    2.25

     Cement mortar (1:2) (as Example 6.61)   0.08
     Gaskin (allow)   0.40
                                                    ─────
                                                    2.73
     Allow for waste (5%)   0.14
                                                    ─────
                                                    2.87

(iii) Labour
      Laying   0.08 gang hour
      Jointing 0.08 gang hour
                ─────
                0.16 gang hour @ £9.58/hour   1.53
                                                    ─────
                                                    4.40
                                                    ═════

Net unit rate per Nr   **£4.40**

![Example 6.63]

Extra; branches	Nr

DATA
(i)   Gang ratio – 1 pipelayer to 1 labourer
(ii) Basic rates:
      150 mm diameter branch   £17.50 each delivered
      150 mm diameter clay pipe   £8.75 per piece
         (1.60 m) delivered
      Pipelayer   £4.86/hour
      Labourer   £4.72/hour

SYNTHESIS	£	£
(i) Materials		
150 mm branch @ £17.50 each delivered	17.50	
Unload (allow)	0.10	
	17.60	
Adjustment for branch being extra over price		
Deduct 1 metre 150 mm pipe @ £ 5.19/m	5.19	
(as Example 6.62)		
	12.41	
Cement mortar (1:2)		
(as Example 6.61) 2 joints @ £0.08 each x 2	0.16	
Gaskin for 2 joints (allow)	0.80	
	13.37	
Allow for waste (5%)	0.67	
	14.04	
(ii) Labour		
Lay branch	0.08 gang hour	
Jointing 2 number	0.16 gang hour	
	0.24 gang hour @ £9.58/hour	2.30
	16.34	

Net unit rate per Nr	**£16.34**

CHAPTER

7

COMPLETING THE TENDER

GENERAL

After pricing all items in the measured sections of the bill of quantities, the estimator will need to price the preliminary items. He or she will also prepare the report which will embrace all facts and assumptions on which the estimate is based.

PRELIMINARIES

Preliminaries are the contractor's project overheads, which fluctuate from project to project. They comprise items of significant cost which, because of their relevance to many items (in lieu of any one particular item of work such as scaffolding), cannot be properly included in the unit rates. Hence preliminary items are priced separately and individually on their merits to reflect the requirements of the project.

As the size and complexity of a project, its value, location, contract conditions, accessibility, and so on, influence the extent of preliminaries, these items may be expediently priced at the following stages.

1. After a site visit and familiarizing oneself with site conditions.
2. After establishing the full extent of the work.
3. After determining the programme time for completing the work.
4. After clearing any outstanding queries with the architect.
5. After pricing the measured items in the bill of quantities.
6. After assessing the subcontractors' requirements.

The Standard Method of Measurement (7th edition) and the *Standard Forms of Contract* provide general information, specific conditions and requirements for preliminaries to be included in the bill of quantities. At the same time, the estimator, in pricing the items, should also include any *additional* items of cost that he or she considers would be best priced in the preliminaries section of the bill.

FIXED AND TIME-RELATED CHARGES

Under Section A of the SMM7, definition rules D1 and D2, the contractor is required to classify his or her preliminaries cost either as a fixed charge or as a time-related charge, as follows:

1. *Fixed charge* A fixed charge, according to the SMM7, covers items of work the total cost of which is neither influenced by time nor directly relates to quantities of items of work. It is a cost that is incurred as a result of the occurrence of an event such as the cost of setting up site facilities or cleaning the works on completion.
2. *Time-related charges* Time-related charges are for items of work total cost of which is dependent on duration of maintaining site facilities or operating major plant on site. The longer the duration of maintenance of these preliminary items, the higher the cost.

The classification is aimed at minimizing disagreements which arise during construction in the identification of charges either as fixed or time-related. The advantages and disadvantages of this classification may be summarized as follows.

Advantages

☐   Easy assessment of preliminary items for interim valuations, thus effecting savings in administrative effort and time.
☐   The contractor is constantly aware of the remaining amount of cash for preliminaries to be released to him or her periodically in interim valuations.
☐   It facilitates the assessment of money that may be due to the contractor for maintenance of site facilities and so forth if the contract period is extended due to the disruptive actions of the client.
☐   It may result in higher payment being made to the contractor at the beginning to the contract, thus assisting him or her in the maintenance of a healthy cash flow.

Disadvantages

○ The contractor may deliberately price his or her fixed cost for items required at the beginning of the contract exceptionally high. This may lead to a high interest charges liability to the client as a result of high initial expenditure.
○ The separation of cost of preliminary items into fixed and time-related cost is somewhat academic as, due to pressure of work, estimators may be reluctant to abide by the rules.

While the above classification is useful, most cost items in the preliminaries do not fit easily into it as most items contain an element of each classification. For instance the cost of a site hut contains the following elements of costs: providing; setting up; maintaining; removal; and making good after removal. Therefore it is incumbent on the estimator to assess the cost of each element, thus enabling him or her to arrange these elements to fit the classification.

For example, looking again at the cost elements of the site hut, only *maintaining* is a time-related charge; *providing, setting up, removal* and *making good* are all fixed charges. It is therefore up to the estimator to examine each cost item in the preliminaries and where appropriate treat each in a similar way. Example 7.1 shows in detail a typical set of preliminary items usually priced by the estimator .

Example 7.1

A set of priced preliminary items based on an estimate for the construction of a two-storey office building with a tender price of approximately £1 million and a contract period of 50 weeks.

Preliminaries

	Fixed charges £	Time-related charges £
Site management staff		
Contracts manager £120.00 per week = 50 weeks x £120.00		6 000
Site agent £450.00 per week = 50 weeks x £450.00		22 500
Project surveyor £240.00 per week = 50 weeks x £240.00		12 000
Site engineer £425.00 per week = 50 weeks x £425.00		21 250
Temporary buildings		
Site office 50 weeks x £55.00		2 750
Mess hut 50 weeks x £45.00		2 250
Storage shed 50 weeks x £15.00		750
Temporary plumbing, drainage and electrical	600	
Temporary fencing		
Fixing and removing	1 000	
Maintenance 50 weeks x £5.00		250
Temporary services		
Water		
Connection (including Water Board charges)	500	
Temporary plumbing	400	
Water for the works 50 weeks x £15.00		750
Electricity		
Connection	400	
Temporary installation	600	400
Cost of electricity 50 weeks x £45.00		2 250
Heating and lighting offices, etc.		
50 weeks x £20.00		1 000
Carried Forward	3 500	72 150

Preliminaries (continued)

		Fixed charges £	Time-related charges £
Brought Forward		**3 500**	**72 150**
Telephones			
Installation		250	
Rental	4 quarters x £ 35.00		140
Calls	4 quarters x £225.00		900
Temporary road and hardstandings			
Provision and removal		3 500	
Maintenance	50 weeks x £10.00		500
Safety, health and welfare			
Site toilets (including toilet material)	50 weeks x £15.00		750
Sewer connection		200	
Temporary plumbing and removal		500	
First aid			150
Protective clothing			500
Scaffolding			
Provision, erection and dismantling		3 500	
Hire cost	50 weeks x £35.00		1 750
Drying out			
Heaters and fuel			300
Site Security	50 weeks x £50.00		2 500
Carried Forward		**11 450**	**79 640**

Preliminaries (continued)

	Fixed charges £	Time-related charges £
Brought Forward	11 450	79 640

Site cleaning and disposal of debris/rubbish

		Fixed charges £	Time-related charges £
Intermittent cleaning	50 weeks x £25.00		1 250
Final cleaning		200	

Sampling and testing

		Fixed charges £	Time-related charges £
Concrete cubes	50 number @ £ 25.00	1 250	
Brickwork panels	2 number @ £120.00	240	

Travelling expenses
Assumed £2.00 per day

		Fixed charges £	Time-related charges £
for 1 operative	50 weeks x 5 days x £2.00		500

Plant and small tools

		Fixed charges £	Time-related charges £
Concrete mixers*	2 number		
Hoist	2 number		
Compressors	50 weeks x £550.00		27 500
Generators			
Fork-lift			

Site transport

		Fixed charges £	Time-related charges £
Tractor and trailer	50 weeks x £35.00		1 750

Alternatively this cost can be allowed for in the unit rates for measured works.

	Fixed charges £	Time-related charges £
Carried Forward	13 140	110 640

Preliminaries (continued)

	Fixed charges £	Time-related charges £
Brought Forward	13 140	110 640

Insurance of the Works

	£		
Contract value	1 000 000.00		
Estimated increased costs			
(a) Contract period (say) 5%	50 000.00		
	£ 1 050 000.00		
(b) Reinstatement period (say) 7%	78 750.00		
Allowance for demolition	25 000.00		
	1 153 750.00		
Fees @ 16%	184 600.00		
Allow 0.15 %	£1 338 350.00	2 008	
		£15 148	£110 640

	£
Total cost of preliminaries	
Fixed time charges	15 148.00
Time-related charges	110 640.00
To tender summary	£125 788.00

PREPARATION OF REPORT FOR MANAGEMEI

On completion of the estimate, and after careful review to ensure that all rates are accurate and no key points have been overlooked in the pricing, the method statement or the programme, the estimator prepares a tender report for consideration by management.

Basically this report should describe the project and the construction method adopted in the preparation of the estimate, and should detail risks and unresolved technical and contractual problems. Information required to be given in the report may be summarized as follows:

1. Description of project, form of construction and any special features.
2. Risks which lack adequate coverage in the contract documentation.
3. Unfamiliar or outstanding technical/contractual problems.
4. State of the design development and consequential productivity implications.
5. Construction method on which the estimate is based.
6. Major assumptions on which the estimating decisions rest.
7. Terms of domestic subcontractor quotations, qualifications and non-standard or unfamiliar conditions.
8. List of items to be considered in relation to possible qualification of the bid.
9. Level of profitability of the project.
10. Winter working and its effect on productivity.
11. Level of retention.
12. State of the construction market and industrial climate pertaining to the project.
13. Summary of the estimate and site visit report.
14. Statement detailing 'firm price risk calculation' where tender is required to be on a 'firm price' basis.

THE TENDER

As contractors are price takers and, at the same time, compete for projects, it is essential that a good decision on which the success of the tender rests (the level of mark-up that will win the contract at an optimum profit level) is made. This is a delicate decision which involves the study of market forces, political-cycle predictions and economic trends, analysis of the competitor factor, competitive advantage, and tender success ratio. The time for this delicate decision arrives when the management meet to consider the estimate in the light of the tender report. This process is known as adjudication, and it is a commercial function based on the estimate.

During the adjudication, management study and analyse carefully all substantive items in the estimator's report. They weigh this information against the following factors to strike the right balance which will either win or lose the contract.

THE REPUTATION OF THE CLIENT

A reputable client may be one whose projects are prestigious and would enhance the contractor's market standing if the project is won. It may also be that the client has a programme of developments in which successful completion of one may lead to commissions for more. If, in management's view, the client is one they would like to do business with, then they will do their best to lower the profit mark up and establishment charges to win the project.

EXPERIENCE WITH THE CLIENT AND HIS OR HER PROFESSIONAL ADVISERS

If the contractor has had previous dealings with the client and his or her professional advisers, then he or she may be in a better position to predict the level of risks that may confront him or her. If experience has shown that the client does not pay promptly and his or her professional advisers are noted for the late issue of project information or disruptive variations, the contractor may decide to increase the project overhead costs to mitigate this risk.

STATE OF THE ORDER BOOK

If the contractor's order book is virtually empty then he or she will be more than willing to cut down his or her profit mark-up and win the project. This move should enable him or her to achieve his or her turnover target still at a profit and also keep key personnel in employment.

If on the other hand, the order book is reasonably full, the contractor would only overcommit him- or herself for a handsome return. In that case, he or she will increase profit mark-up to enable him or her to finance additional key personnel that he or she will have to acquire if the bid is successful.

CAPITAL REQUIRED FOR THE PROJECT

This is the contractor's evaluation of the magnitude of extra capital he or she may require to finance the project. In this assessment he or she will consider the following factors.

Work in progress

When work in progress is likely to be completed so as to release money tied up for the proposed project.

Investment in plant

If the project requires the use of specific plant which can be fully utilized to the contractor's advantage, can he or she conveniently tie up part of his or her working capital in outright plant purchase? If not, then is he or she in the position to raise the loan required to purchase plant and undertake the project economically?

Level of retention money

Retention money is a percentage deduction of money due to the contractor to safeguard the client's interest. This deduction means that the contractor cannot put all his or her working capital to work and has to rely on borrowed funds to make up the shortfall. If the level of retention is high, then the contractor will have a greater proportion of his or her working capital tied up in retention, which may result in uneconomic borrowing to finance the project.

Technical and managerial manpower requirement

If the complexity of the project requires more technical and managerial input than the current resources at the contractor's disposal, he or she may consider either employing or hiring consultants to undertake some of the technical and managerial functions. If this should be the case, then the question of the availability of these key personnel and their cost of engagement arises. He or she may even decide to train some of his or her own staff to deputize for some of these functions in the short term.

STATE OF THE ECONOMY AND ITS EFFECTS ON THE CONSTRUCTION MARKET

As construction is not a leading sector of the economy, it tends to follow the up-and-down movements of other sectors, the general national and international economies. A study of the economic indicators will enable the contractor to forecast whether the economy is heading for a boom or recession. If the indications are that recession is imminent, which may slow down construction activities, then the contractor would price keenly to win the contract. Determination to win means lower percentage mark-up and hence reduced profit margin. However, the reverse will be the case should the indicators show pending boom in the national economy.

CONTRACTUAL RISKS

In the construction industry, contractors are required to assess cost and price of a product before production, and this process involves high risks due to the uncertainties of the national construction market, the national and international economies, the weather, ground conditions, and so on. The risks to be considered by management may include the following factors.

Contractor's resources

1. *Capital* The cost of borrowing may rise above the anticipated rate of interest and capital may not be utilized efficiently due to irregular payments, undervaluations for interim payment and construction delays.
2. *Materials* Material prices may change over time due to acute shortages, high interest rates or a fall in the value of the local currency. Late deliveries may lead to uneconomic working and a possible overrun on the contract programme.
3. *Labour* The site may be plagued by labour shortage, strikes, low morale, go-slows and low productivity. An untried labour force could result in faulty workmanship which could be difficult to rectify.

Weather conditions

Adverse weather conditions are unpredictable and this can affect ground conditions and disrupt progress on site.

The client and his or her professional advisers

They may not supply project information on time, leading to an extended contract programme and additional costs that may be difficult to substantiate and may result in financial loss.

The estimate

This may be inaccurate due to either the fact that a large proportion of the data used in the preparation of the estimate is outside the contractor's organization (e.g. domestic subcontractors' and suppliers' prices) or a proportion of the project may be outside the estimator's and management's experience and hence they cannot assess the full cost implications accurately.

MARKETING STRATEGY

Management may plan for business expansion and so require new clients, or the project may be one which suits the firm's operational strategy. If business expansion as opposed to business contraction is management's current policy, this strategic consideration will influence their decision to submit a more competitive bid.

RE-EXAMINATION OF THE OUTLINE PROPOSAL

Management will re-examine the outline construction programme and construction methods to see if there is room for improvements that might lead to the reduction of cost and/or general overheads.

After considering all these factors, management will be in a better position to arrive at a competitive percentage mark-up to be added to the estimated cost of the project. This figure will be based on the framework of the company's tendering policy, management's assessment of market conditions, levels of competitive advantage, and analysis of the firm's historical tender performance in relation to that of its competitors. If the estimate is accurate enough, then this figure should stand a better chance of winning the contract with minimum risk and, hopefully, make a healthy contribution towards the contractor's overheads and profit.

ADJUSTMENT TO NET ESTIMATE

Assuming that the estimator has estimated the net total cost of the project to be £774 712 and the preliminaries £125 788, and management has adjudicated that establishment charges should be 7.55 per cent and profit 4.5 per cent, this may be set out as follows.

Net Estimated Total Cost of Works Contained in the Bill of Quantities

	£	£
Measured work (say)	599 712.00	
Preliminaries	125 788.00	725 500.00
PC and provisional sums (say)		175 000.00
		900 500.00
Profit and establishment charges 12.05% of £725 500.00		87 422.75
		£987 922.75

NET OR GROSS RATE PRICING

The next stage in the estimating process is the estimator's decision on the method of allocating the sum total of preliminaries, establishment charges and profit to the bills of quantities. The method adopted will depend on the preferred method of pricing the bills of quantities and this may be either net rate pricing or gross rate pricing.

Net rate pricing

Where net rate pricing is the preferred method, the unit rate for each measured item contained in the bill of quantities is calculated net, that is, without the overhead charges and profit. A lump sum for overhead charges and profit is added to the sum total of the extended net unit rates to arrive at the tender figure (see Example 7.2).

Alternatively this may be varied to show the sum for preliminaries separately and establishment charges and profit element expressed as a percentage mark-up on measured work (see Example 7.3).

Example 7.2

Summary of tender	

	£
Preliminaries	125 788.00
Measured work	599 712.00
PC and provisional sums	175 000.00
Overhead and profit	87 422.75
To form of tender	£987 922.75

Example 7.3

Calculation of percentage mark-up	

$$\frac{(\text{Establishment charges} + \text{profit})}{\text{Net total of measured work}} \times 100$$

$$\frac{£87\ 422.75}{£599\ 712.00} \times 100 = 14.577455\% \text{ (say) } 14.58\%$$

Summary of Tender

	£
Preliminaries	125 788.00
Measured work	599 712.00
PC and provisional sums	175 000.00
	900 500.00
Percentage mark-up 14.58% of £599 712.00	87 438.01
To form of tender	£987 938.01

Note: The slight difference in total is the result of limiting the percentage addition to two decimal places in Example 7.3.

The 'net rate pricing' method has the advantage that the estimator can calculate and allocate the overhead and profit elements to the bill of quantities separately. However, it has the disadvantage that the overhead and profit elements are not affected automatically when bill prices are used in the valuation of variations. Where establishment charges and profit are hidden in the preliminaries and not shown separately as in Examples 7.2 and 7.3, difficulties are experienced in calculating the correct percentage additions to net rates.

Gross rate pricing

Under this method of pricing, preliminaries are priced separately and individually. Establishment charges and profit are expressed as a percentage of the measured work as in Example 7.3 but the unit rate for each measured item contained in the bill of quantities is increased by this percentage (see Example 7.4). The sum total of the extended gross unit rates becomes the tender figure.

Example 7.4

Gross rate pricing

Calculation of percentage mark-up (as Example 7.3) = 14.58%

Calculation for gross unit rate

	£
Excavation of trench (net unit rate from Example 6.4)	5.36
Add	
Percentage mark-up to cover establishment charges and profit 14.58%	0.78
Gross unit rate per m³	**£6.14**

The advantage claimed for the gross rate pricing method is that overhead and profit elements are automatically affected when bill rates are used in valuation of variation. However, this method has the disadvantage that the adding of percentage mark-up to each net unit rate exposes the estimator to a greater margin of error. Moreover, as compared with net rate pricing, the gross unit rates appear inflated (see Example 7.4), and also more time is consumed in the estimating.

FIRM PRICE TENDER RISK CALCULATIONS

Sometimes contractors are invited to tender on a firm price basis. What this means is that the fluctuations clause is not applicable in the proposed contract and hence contractors cannot recover any increased costs of labour, plant and materials that may occur throughout the duration of the contract. Therefore, when tendering on a firm price basis, the estimator/management is required to forecast all future price increases on resources (i.e. labour, plant and materials) and allow for them either in the unit rate build-up or when finalizing the tender.

The risk in a firm price tender can be great, especially in periods of high levels of interest rates and inflation. The estimator and management (at adjudication) are faced with the daunting task of predicting the likely changes in the national and international economies in the short term, which might affect the cost of resources to be employed in the execution of the building works.

Generally, projects for which firm price tenders are invited are of a short duration. Normally they range from 12 to 18 months. However, no matter how short the contract period is, material suppliers, plant hire companies and site operatives, as a general rule, will not observe and abide by the terms of any fixed price arrangement. Therefore the estimator and management need to study carefully the likely path the economic variables (e.g. inflation, interest rates) will take in the future and to make adequate provisions in the tender to cover them. Moreover they will need to analyse the current national and international economic variables and project them into the future by using historic data.

INDEX NUMBERS

An index measures changes from one period to another and it is one of the most important factors of forecasting techniques that rely on historic data. Therefore, in the construction industry, construction cost index measures change in the cost of a construction item or a group of items from one point in time to another. As the cost of construction is determined primarily by the cost of resources (i.e. labour, plant and materials) involved in its execution, a variation in any of these factors will influence the total construction cost. Thus there are available for use several cost indices that are intended to provide an empirical guide to movement of cost of construction resources and activities. These indices are compiled by official bodies (e.g. Department of the Environment, *Bulletin of Construction Statistics*, Royal Institution of Chartered Surveyors, *Building Cost Information Services (BCIS) and Building Maintenance Cost Information Service*) and are updated from time to time. By plotting the pattern of construction costs, the estimator and management are able to extrapolate the trend of these costs into the short-term future with some confidence.

TYPES OF INDICES

As mentioned above, there are a diverse number of cost and price indices available for use in the construction industry. However, the following two are relevant to our exposition:

1. *Building cost indices*   Building costs refer to the construction costs actually incurred by the contractor in carrying out the building works. These costs include factors such as labour, material and plant costs, rates, rents, overheads and taxes. The indices in this respect measure movements in these factor costs over a given period (e.g. 3 months, 12 months).
2. *Tender price indices*   Tender price is the price a client pays to the contractor for executing the building works and includes building cost plus other considerations such as:
   (a)   contractor's profits; and
   (b)   the contractor's anticipated cost during the lifetime of the building contract.
   Hence tender price indices measure the level of prices charged to clients by contractors over a given period.

The above indices are obtained from the *Building Cost Information Services (BCIS)* compiled and published by the Royal Institution of Chartered Surveyors (RICS). The RICS also compiles and publishes a number of other building cost information papers for use in the construction industry.

ADJUSTMENT FOR FIRM PRICE TENDER

The anticipated increases in construction costs are determined by the use of cost indices. The estimator/management with the aid of indices can therefore calculate the increase on either the total construction costs or on the elemental cost of resources.

Increased total construction cost

The determination of the predicted increases in total construction costs employs the use of BCIS *Total Construction Costs Index* and the formula for calculating percentage change as follows:

$$\frac{It - Io}{Io} \times 100$$

Where: Io = base index
              It = the finishing index for calculating increased cost

With this the project data is as follows:

1. *Value of measured works (gross)*

Total tender cost	£987 922.75
Less	
PC and provisional sums	£175 000.00
Measured work (including 12.05% profit and overhead)	£812 922.75

2. *Contract period*

    Given contract period = 50 weeks

    Assumed tender date:
    June 1993 = index base date

    Assumed project completion date:
    May 1994 = index completion date

3. *Cost indices*

    Percentage change from June 1993 to May 1994:
    Second quarter of 1993 = 149
    Second quarter of 1994 = 152 (extrapolated)

    Therefore percentage change:
    $$= \frac{152 - 149}{149} \times \frac{100}{1}$$
    $$= \frac{3}{149} \times \frac{100}{1}$$
    $$= 2\%$$

The 2 per cent anticipated increase represents an increased cost at the end of the contract period. However, as the construction cost is spread over the entire contract period and periodic interim payments will be made (and will not be paid as a lump sum at the end of the period), it is therefore prudent to use half-point value, that is, 1 per cent of the anticipated percentage increase.

4.   *Increase cost allowance for tender*

Anticipated value of increase:
1% of the builder's work £812 922.75 = £8 129.23

Depending on how the estimator intends to present the tender, this anticipated increase can be built into the preliminaries. Alternatively, it can be added to the lump sum profit or expressed as a percentage of the measured work and the unit rates adjusted accordingly.

Increased elemental costs of resources

This method utilizes BCIS *Factor Element Cost Index*. The predicted increases in the cost of labour, materials and plant can be determined by applying the following input indices: basic labour cost; basic materials cost; and basic plant cost. With this method, the data is as follows:

1.   *Calculating the increased cost of labour*

Second quarter of 1993   Index = 161
Second quarter of 1994   Index = 162

Using the formula: $\dfrac{It - Io}{Io} \times 100$

$$\text{Uplift} = \frac{162 - 161}{161} \times \frac{100}{1} = 0.62\%$$

As explained above, the 0.62 per cent represents an increased labour cost at the end of the contract period and hence the half-point average value will be:

$$\frac{0.62}{2} = 0.31\%$$

2.   Calculating the increased cost of materials

Second quarter of 1993   Index = 139
Second quarter of 1994   Index = 143

$$\text{Uplift} = \frac{143 - 139}{139} \times \frac{100}{1} = 2.88\%$$

As explained above, the half point value will be:

$$\frac{2.88}{2} = 1.44\%$$

3.   *Calculating the increased cost of plant*

Based on index for the period between June 1992 to May 1993 (due to unavailability of current index) the following assumptions have been made:

Second quarter of 1993   Index = 143
Second quarter of 1992   Index = 136

For June 1992 to May 1993:

$$\text{Uplift} = \frac{143 - 136}{136} \times \frac{100}{1} = 5.15\%$$

With the increase for the period 1992–1993 calculated and taking the key economic indicators into account, a hypothetical increase of 35 per cent may be predicted for the 1993–1994 period in our example. The predicted increase can therefore, be calculated as follows:

5.15% + 35% of 5.15% = 5.15 + 1.80 = 6.95%

As explained above, the half point value will be:

$$\frac{6.95}{2} = 3.48\%$$

4.   *Increase cost allowance for tender*

The above calculations show that during the contract period the various resources will increase as follows:

Labour	0.31%
Materials	1.44%
Plant	3.48%

Therefore based on the above calculations, the estimator would increase the labour rate by 0.31 per cent, material price by 1.44 per cent and plant rates by 3.48 per cent in the unit rate build-up to cover for the likely future price increases.

EFFECT OF VALUE ADDED TAX (VAT) ON TENDERS

Value Added Tax (VAT) is a tax levied on goods or services an individual or organization procures (i.e. consumer expenditure) and it is administered by Her Majesty's Customs and Excise. Whilst it is in operation every business with a certain minimum taxable turnover is required to register for VAT. On registration, the business must then invoice and collect the tax for the goods and services it performs at the appropriate rate.

The VAT assessment period normally occurs every three months and this means that businesses make returns every three months. However, by a special arrangement, a business may be permitted to submit returns either monthly or annually. During the assessment, the differences between VAT paid by an organization for goods and services procured (input VAT) and that received for the goods and services supplied (output VAT) is either paid to or collected from HM Customs and Excise.

VAT AND THE BUILDER'S TENDER

The builder normally bases his or her tender on materials, plant and other services that are subject to the VAT provisions. Nevertheless VAT is not normally included in the tender sum as builders recover all input VAT paid quarterly from HM Customs and Excise.

During the progress of the work, the client pays the builder VAT on all net interim certificates as well as the final certificate issued by the architect. The builder, in turn, issues the client with an authenticated VAT receipt and pays the tax to HM Customs and Excise. If the client is registered for VAT, he or she can then claim the tax paid from HM Customs and Excise; otherwise he or she absorbs the tax burden.

CHAPTER

8

TENDER ANALYSIS

TENDER CHECK

When tenders have been returned and opened the two lowest tenderers are, in practice, asked to submit their priced bill of quantities for examination. This request becomes necessary where tenderers have not been instructed to return the priced bill of quantities with their bids. The examination of the priced bill of quantities is conducted by the client's quantity surveyor who checks the following points in each bill:

1. *Bill pages*   All pages in the bill of quantities are intact and all items have been priced.
2. *Unpriced items*   A short line is drawn through the cash column of items the tenderer did not intend to price.
3. *Alterations*   The bill of quantities contains no unauthorized alterations.
4. *Amendments*   All notified corrections have been made in the relevant parts of the bill of quantities.
5. *Correct figures*   The bill extension, casts and collections are arithmetically correct and have been transferred to the summary page and form of tender.
6. *Correspondence of figures*   The figure in the summary page of the bill of quantities agrees with the tender figure.
7. *Fair pricing*   The pricing is satisfactory and uniform throughout, and the unit rates could be used as a fair basis for the valuation of variations.
8. *Consistency of pricing*   The prices for the same items of work in different sections of the bill of quantities are consistent.
9. *PC and provisional sums*   Reasonable sums for percentage and

      lump sum addition for profit and for attendance on specialist subcontractors and suppliers have been made.

10. *Dayworks* Percentage additions made to the prime cost of daywork items (labour, plant and materials) are acceptable.

11. *Preliminaries* Preliminaries section of the bill of quantities has been priced individually, and fixed and time-related charges are clearly shown.

12. *Serious mistakes* The tender contains no serious mistakes that will force the tenderer to withdraw his or her tender.

13. *Qualifications* The tenderer has not made any qualifications to his or her tender.

The technical check above is aimed to ensure that all anomalies are clarified and agreed before the client is committed to contract. Failure to do so creates problems and confusion once work has started on site.

 ## CORRECTION OF ERRORS

A tender sum is a legal offer made by a contractor to carry out work and hence is not subject to adjustment. This figure will only change by reason of variations to the contract. Example 8.1 is an example of how arithmetical errors are corrected in a priced bill of quantities.

Example 8.1

Correction of error in a priced bill of quantities

1. *Tender Sum*

	£
Builder's work	687 134.75
Preliminaries	125 788.00
PC and provisional sums	175 000.00
	£ 987 922.75

However, corrected errors make this £978 912.50

2. *Revised value of builder's work*

	£
Corrected Tender	978 912.50

Less

Preliminaries	125 788.00	
PC and provisional sums	175 000.00	
		300 788.00
Value of builder's work		£678 124.50

The corrected builder's work is less than it should be by £9,010.25 therefore the following error adjustment is required:

3. *Error Adjustment*

$$\frac{\text{Error}}{\text{Corrected value of builder's work}} = \frac{9\,010.25}{678\,124.50} \times 100 = 1.3287\%$$

Therefore to arrive at the correct tender sum, all unit rates in that builder's work are subject to 1.3287% addition. This may be demonstrated as follows:

	£
Corrected value of builder's work	678 124.50
Add 1.3287% =	9 010.24
	687 134.74
Add sum of preliminaries, PC and provisional sums	300 788.00
	£987 922.74

The above calculations mean that, during the production phase, all builder's work included in the interim valuations and any variation valued at the bill rates are subject to an additional adjustment of 1.3287%.

Note: The difference of £0.01 in the above example is the result of limiting the percentage addition to four decimal places.

VARIATIONS IN TENDERS

Although the same project information is distributed to competing contractors, the client receives bids normally within the ranges of ±10 per cent of the estimated cost of the project. The reasons for these differences may be legal, commercial, technical, administrative, economic, political, and human/social circumstances, and may vary in their individual and collective influence upon the contractor. The following factors may be considered as the reasons for differences in tenders submitted by contractors for projects.

Project appreciation

The method of calculating the cost of the construction works is influenced by the estimator's interpretation of the information contained in the tender documents. Where the estimator is under a false impression that the contract contains onerous conditions and the standard of workmanship required is higher than average, his or her method of pricing will differ significantly from someone who fully appreciates the requirements of the project.

Contractor's past experience with client and his or her advisers

The contractor's previous experience with a client and/or his or her professional advisers influences the estimator's pricing method. Where the client is known to make late or irregular payments an allowance is made in the pricing to cover the extra cost.

Differences in quotations and prices

Contractors work with different subcontractors and plant hire firms and hence receive different prices from these firms. This eventually leads to differences in pricing the measured works when these prices are used in unit rates build-up.

Purchasing arrangements

Variations in cost of materials for a project may arise from the following factors:

1. The skill and experience of the contractor's buyers may lead to procurement from a reliable source of supply on favourable terms.

2. Frequency of purchase and large orders attract discounts and other facilities (e.g. reduced cost of delivery).
3. Prompt settlement of credit account on demand enables the contractor to enjoy credit facilities.

Contractor's business organization

Contractors who own plant departments or subsidiary companies are able to achieve lower costs for the supply of services and equipment. This association will enable the contractor to procure goods and services at a cheaper rate than his or her competitors.

Method statement and construction programme

The estimator's method of pricing is influenced greatly by the method statement and construction programme. As the methods adopted vary, so do the construction programmes for various activities and their associated costs.

Operational efficiency

Contractors have different levels of cost which reflect their operational efficiency. An efficient firm has a lower operating cost and therefore is able to price work more competitively.

Quality of estimating

1. *Experience of estimator* The skill and experience of estimators vary and an experienced estimator is in a better position to give a reasonably accurate price than one without much experience.
2. *Errors* Estimators do make errors during the preparation of the estimates, which may be due to a variety of reasons (e.g. wrong assessment of local customs and conditions, non-availability of labour of the right quality, wrong establishment of labour constants, misinterpretation of descriptions and taking wrong measurements where the estimate is based on drawings and specifications); hence different levels of pricing.
3. *Waste factors* Various allowances are made for waste in material usage. Estimators vary in making this allowance and this leads to different unit rates.
4. *Project and head office overheads* Contractors have different overheads which arise from diverse staffing levels, salaries, di-

rectors' fees, and so on, and the magnitude of these costs is reflected in the estimator's pricing.

5. *Adverse weather*   Estimators make dissimilar assessment of the cost of winter working and their pricing methods are affected by these assessments.

6. *Travelling allowance*   The level of travelling allowance included in estimates depends on where the operatives live in relation to the location of the project. As this varies from firm to firm, it is priced differently and is reflected in pricing levels.

Fluctuation

On fixed price contracts, estimators are required to make predictions on the probable increase in the cost of resources throughout the duration of the contract. These predictions are bound to differ and hence provoke differences in pricing levels.

Adjudication

Various adjustments are made to the estimates by management at the adjudication stage to reflect the company's current workload, competitive advantage, tendering policy, and known contractual risks and uncertainty. All these factors vary from firm to firm and in addition the contractor considers the following points:

1. The required levels of mark-up for overhead and profit. Where a contractor is keen to win the contract, he or she will trim the mark-up to reflect this desire.

2. Prediction of level of errors in the estimate; the higher the level of suspected errors, the higher will be the addition to cover the risk.

3. Prediction of level of inflation and interest rates. This is a difficult prediction to make as economies of nations are interdependent, and allowance made by contractors to cover these risks will vary.

4. Uncertainty about the possibility of overrunning the contract programme and allowance made to cover the damages associated with this risk.

DIRECTION OF TRADITIONAL TENDERING

The foregoing has thrown some light on building contract procurement under the traditional system which operates on the separation of design and production phases. However, the circumstances of building production have changed in recent years in such a way that the traditional tendering procedures no longer adequately reflect the requirements of a modern commercial/industrial client. Although it may be said that the traditional contractual procurement arrangement may be adapted to meet the rapidly developing situations, these are, however, fraught with difficulties. The reason for this is that factors which influence the adaptation (e.g. clients' requirements, project particulars and circumstances, the expertise, ability and capacity of firms in the industry, environmental constraints) do not facilitate adaptation. Hence there is a declining use of the traditional contractual procurement, influenced by the following factors.

Increased technological complexity

Modern building has become complex in terms of size, diverse component parts, an increased amount of services element, and technically specialized parts. This necessitates input by specialist designers and contractors during both design and production phases. However, this early input by a contractor is not suitable for the contractual procurement arrangement where design is separated from production.

Expedited construction programme

Corporate clients demand their buildings earlier than before. This circumstance has led to an early contractor involvement during the design phase to advise on buildability, availability of resources and contentious details that may hold up the production process on site. This approach requires a suitable mode of contractor selection and legal arrangement in a situation of dual responsibility for design.

Design and construct

These days most contractors are well equipped and offer the services of design as well as construction. This combined service approach does away with the traditional tendering arrangement described above.

Construction management

The activities in construction have become dispersed among many specialist works contractors. Therefore, the main contractor of today hardly takes part in direct execution of the works. Rather he or she organizes and co-ordinates the input of works contractors to whom all the works are sublet. A revised tendering arrangement is emerging to reflect the main contractor's co-ordinating function and the important role played by the works contractors.

Negotiations

Many clients have successfully completed many projects through negotiated tender. In this approach, the client negotiates with the contractor on matters of mutual advantage to their respective business rather than on cost alone.

Less detailed sets of contract drawings

Clients recognize that frequent litigation arises largely from inadequate contract documentation. Contractors are now required to assume the responsibility for design details and give an undertaking of adequacy of design information prior to contract (i.e. the BPF system). This system makes new tendering arrangements necessary to reflect the contractor's newly assumed detailed design responsibility.

While the above factors may have contributed to some modifications in traditional tendering arrangement, they have had no effect on tendering cost, which still constitutes a major item of cost to clients of the industry. Moreover the advent of non-traditional construction procurement methods explained above has yet to have any cost reducing effect on tendering. Tendering cost under the contemporary tendering methods is passed down to works contractors who bid for the various work packages. Hence the identification of a simple and economical method of contractor selection will be beneficial to the construction industry as a whole.

TENDERING ARRANGEMENTS OUTSIDE THE UNITED KINGDOM

This section is intended primarily to give a basic awareness of tendering practices and procedures of countries outside the United Kingdom. It is hoped that it will enable readers to draw comparisons of the other systems with that practised in the United Kingdom. The countries whose systems are considered include Belgium, Denmark, France, Germany, Greece, Italy, the Netherlands, United States of America, the Middle East and Commonwealth countries.

BELGIUM

Design
The design of buildings is carried out by a university-trained registered architect who signs drawings and gives a ten-year guarantee.

Types of contract
Contracts in regular use are:

1. A strictly fixed lump sum contract *(Marché à forfait absolu)* with no provision for variations in quantities or contract sum.
2. A lump sum contract *(Marché à forfait relatif)* with provisions for both variations and fluctuations.
3. The remeasurement contract *(Marché à borderaux de prix)* is based upon a schedule of rates.

Methods of obtaining tenders
Methods of tendering adopted include:

1. Selective tendering. Used for private work.
2. Negotiated tender. Used for private work.
3. Public tendering. Used for government contracts.

Tender documents

The tender documents are prepared by the architect or the engineer and are composed of drawings, form of contract, general specification for materials and workmanship, and a particular specification.

Tenderer/contractor

The bidding contractor will submit a lump sum price for executing the work and, in a lump sum contract, he or she supports his or her tender with a schedule of quantities and rates which when successful will form the basis of the valuation of variations.

During the progress of the works, periodic net payments (i.e. less 10 per cent retention) based on stages of work completed are made, and the retention money is paid on satisfactory completion of the works.

DENMARK

Design

The design of buildings is carried out by a university trained, registered architect.

Types of Contract

Contracts in regular use are:

1.   Cost reimbursement contract
2.   Lump sum contract

Methods of obtaining tenders

Methods of tendering adopted include:

1.   Open tendering. Used for most government projects.
2.   Selective tendering. Used for government projects.
3.   Negotiated tender. Used for private work.
4.   Package deal tender. Used for private work.

Tender documents

The tender documents are prepared by the architect or the engineer and are composed of drawings, detailed specification, form of agreement, form of tender and general conditions of contract.

Tenderer/contractor

The bidding contractor prepares his or her own schedule of quantities from which he or she submits a lump sum price for executing the works. If successful, and whilst on the job, the contractor submits periodic interim payments for 90 per cent of the work executed. The balance of payment is released on satisfactory completion of the works.

FRANCE

Design

The design of buildings is carried out by a registered or trained architect who is obliged to insure his or her design for ten years against his or her negligence.

Types of contract

Contracts in regular use are:

1. Lump sum contract (with provisions for variations and fluctuations).
2. Package deal contract.

Methods of obtaining tenders

Methods of tendering adopted include:

1. Open tendering *(l'adjudication)*. This is issued for government projects.
2. Selective tendering *(l'adjudication)*. This is also used for government projects.
3. Negotiated tender. Used for private work.

Tender documents

The tender documents are prepared under the direction of either the architect or the engineer and are composed of drawings, specification, national codes of practice, contract conditions, contract agreement and form of tender.

Tenderer/contractor

The bidding contractor prepares his or her own schedule of quantities and submits a lump sum price for executing the works. He or she is also obliged to support his or her tender with a detailed breakdown of the tender as well as supplying rates for valuing variations should he or she be successful.

Whilst on site, he or she periodically prepares interim applications for payment of 90 per cent of the works executed. The balance of his or her money is released on satisfactory completion of the works.

GERMANY

Design

The design of buildings is carried out by a registered architect who is required to take out a professional indemnity policy.

Types of contract

Contracts in regular use are:

1. Measurement contract.
2. Fixed lump sum contract (which allows rates alteration to reflect variation of quantities in excess of 10 per cent).
3. Package deal contract.

Methods of obtaining tenders

Tendering procedure is laid down in *Verdingungsordung für Banleistungen (VOB)* and the method of tendering adopted includes:

1. Open tendering. Used for government projects.
2. Selective tendering. Used for private works.
3. Negotiated tender. Used for private works.

Tender documents

The tender documents are prepared by the architect or the engineer, and they comprise VOB. The latter is in three parts: (1) general conditions; (2) general procedures; (3) general technical regulations, drawings and an approximate schedule of quantities.

Tenderer/contractor

The bidding contractor prices the approximate schedule of quantities and, if successful, an interim payment equivalent to 95 per cent of work completed is paid to him or her periodically during the progress of the works. The sum retained is released on satisfactory completion of the works.

GREECE

Design

The design of buildings is carried out by a qualified architect.

Type of contract

The contract in regular use is a lump sum contract (with provision for both variations and fluctuations).

Methods of obtaining tenders

Methods of tendering adopted include:

1. Open tendering. Used for government projects.
2. Negotiated tender. Used for private projects.

Tender documents

The tender documents are prepared by an architect or an engineer, and usually comprise:

1. Drawings.
2. General specification.
3. Technical specification.
4. Technical description.
5. Work item description and unit price list.

6.   Schedule of quantities and cost estimate.
7.   Form of tender.

Tenderer/contractor

The contractor prices the schedule to arrive at a lump sum price for undertaking the project. If successful, and while the work is in progress, he or she receives a periodic interim payment prepared by an engineer equivalent to 90 per cent of work executed. The retained amount is released to the contractor on satisfactory completion of the works.

ITALY

Design

The design of buildings is carried out by a graduate architect who signs the drawings and has ten years' liability for his or her work.

Type of contract

Contracts in general use are:

1.   Fixed lump sum contract (at times with fluctuation provisions).
2.   Package deal contract.

Methods of obtaining tenders

Methods of tendering adopted include:

1.   Open tendering. Used for government projects.
2.   Selective tendering. Used for government projects.
3.   Negotiated tender. Used for private works.

Tender documents

The tender documents are prepared by an architect or an engineer, and comprise drawings, general form of contract conditions, special specification clauses (applicable to the particular project) and form of tender.

Tenderer/contractor

The bidding contractor prepares his or her own schedule of quantities from which he or she submits his or her lump sum price for carrying out the work. If successful and during the progress of the works, an interim payment equivalent to 90 per cent of the completed work is paid to him or her periodically. The sum retained is released on satisfactory completion of the works.

THE NETHERLANDS

Design

The design of buildings is carried out by an architect.

Types of contracts

Contracts in general use are:

1. Fixed lump sum contract.
2. Lump sum contract (with provision for both variations and fluctuations).

Methods of obtaining tenders

Methods of tendering adopted include:

1. Open tendering. Used for government projects.
2. Selective tendering. Used for private work and government projects.
3. Negotiated tenders. Used for private work.

Tender documents

The tender documents are prepared by an architect or an engineer, and normally include drawings, form of contract, and a general specification for materials and workmanship.

Tenderer/contractor

The bidding contractor prepares his or her own schedule of quantities from which he or she submits a lump sum price for executing the project. He or she is required to support his or her tender with a schedule of quantities and rates in operational order. If successful and during the progress of the works, the contractor receives interim stage payments equivalent to 90 per cent of stage work completed. The retained sum is released to him or her on satisfactory completion of the works.

UNITED STATES OF AMERICA

Design
The design of buildings is carried out by a qualified architect.

Types of contract
Contracts in regular use are:

1. Lump sum contract.
2. Cost reimbursement contract.

Methods of obtaining tenders
Methods of tendering adopted include:

1. Open tendering. Used for government projects.
2. Selective tendering. Used for private projects
3. Negotiated tender. Used for private projects.

Tender documents
The tender contract document is prepared by the architect and includes instructions to bidders, the bid form, the contract, the general conditions, the special conditions, and the specification and drawings.

Tenderer/contractor
The contractor/cost estimator proceeds to work through the above information and prepares a schedule of quantities to enable him or her to obtain an accurate estimate of construction costs.

THE MIDDLE EAST (e.g. SAUDI ARABIA, KUWAIT)

Design

The design of a building is undertaken by a locally trained/qualified or a qualified expatriate architect.

Tendering and contract procedures

The type of contract and methods of obtaining contracts, although locally developed, are based on either *The Conditions of Contract (International) for Civil Engineering Construction (FIDIC)* or *The American Institute of Architects (AIA) Contract Conditions* principles.

Types of contract/tendering

Contracts/tendering methods in general use are:

1. Lump sum contract.
2. Open tendering method. Used for obtaining tenders.

Tender documents

The tender documents are produced by an architect when AIA contract conditions are used or by an engineer when FIDIC contract conditions are in use and comprise the documentation detailed in Table 8.1.

Table 8.1   AIA and FIDIC contract conditions

AIA contract conditions	FIDIC contract conditions
Instruction to bidders	Instructions to tenderer
Bid forms and priced schedule	Form of tender (with appendix)
Agreement form	Conditions of contract
Performance bond form	Specifications
Drawings	Drawings
General conditions of contract	Bill of quantities
Supplementary conditions	Circular letters from consultant engineer
Technical specifications	Geological data

Tenderer/contractor

The contractor prepares his or her lump sum price either by pricing a bill of quantities (project adopting FIDIC contract condition principles) or preparing and pricing a schedule of quantities (projects adopting AIA contract conditions principles) to arrive at a lump sum price for executing the project. If successful and whilst the works are in progress, the contractor receives periodic interim payments.

THE COMMONWEALTH COUNTRIES (e.g. CANADA, AUSTRALIA, GHANA)

The Commonwealth countries are influenced by the United Kingdom's construction practices. The design of buildings in Commonwealth countries are undertaken by a locally/foreign-trained qualified architect or an expatriate architect.

Types of contract

Types of contract in regular use invariably follow some of the contractual arrangement used in the United Kingdom and some of these are as follows:

1. Lump sum contract (with provisions for both fluctuations and variations).
2. Fixed lump sum contract.
3. Measurement contract.
4. Cost reimbursement contract.

Methods of obtaining tenders

Methods of tendering adopted include:

1. Open tendering.
2. Selective tendering (single stage).
3. Negotiated tender.

Tender documents

The tender documents are prepared under the direction of the architect and are composed of the following:

Table 8.2 Summary of contract and tendering methods

Country	Type of contract				Tendering method		
	Lump sum	Measurement	Cost reimbursement	Package deal	Open	Selective	Negotiated
Belgium	✓				✓	✓	✓
Denmark	✓		✓		✓	✓	✓
France	✓				✓	✓	✓
Germany	✓	✓		✓	✓	✓	✓
Greece	✓			✓	✓		✓
Italy	✓				✓	✓	✓
The Netherlands	✓				✓	✓	✓
Middle East	✓			✓	✓		✓
United Kingdom	✓	✓	✓		✓	✓	✓
United States	✓		✓			✓	✓
Commonwealth	✓	✓	✓			✓	✓

1.   Drawings.
2.   Bill of quantities or specification.
3.   Form of tender.
4.   Letter of invitation.
5.   Return envelope.

Tenderer/contractor

The contractor prices the bill of quantities to obtain a lump sum price for executing the project. However, in the case of tenders invited on drawings and specifications only, the contractor prepares and prices a schedule of quantities for the production of his or her tender. The lowest bidder is awarded the contract and while progressing with the project on site, he or she receives a periodic interim payment prepared by a quantity surveyor for work done, with a small percentage, normally 5 per cent, retained until satisfactory completion of the works.

SUMMARY

Table 8.2 is a summary of contract and tendering procedures in the United Kingdom and selected countries.

BIBLIOGRAPHY

GENERAL

Carlvert, R. E. (1990), *Introduction to Building Management*, Newnes.

Coles D. and Smith G. (1981/82), 'Estimating and buying', Building Trade Journal Papers 1–7.

Crawshaw, D. T. (1980), *Project Information at the Pre-Construction Stage*, CIOB Estimating Information Service.

Hillebrandt, P. M. (1974), 'The capacity of the construction industry', *Building*, August, pp. 71–3.

Hughes, W. P. (1992), *An Analysis of Traditional General Contracting*, CIOB Construction Papers.

McCanlis E. W. (1966), *Tendering Procedures and Contractual Arrangements*, RICS Research and Information Group of Quantity Surveyors Committee.

Pilcher, R. (1976), *Principles of construction management*, McGraw-Hill.

Potter, D. G. (1982), *The Training Needs of the Building Estimator*, CIOB Technical Information Service.

Newlove, J. (1979), 'The issue of construction information', *Building Technology and Management*, July/August, pp. 2–5.

CHAPTER 1

Andrews, J. (1989), unpublished college lectures (Construction Management), University College London.

Ball, M. (1988), *Rebuilding Construction*, Routledge.

Hillebrandt, P. M. (1984), *Analysis of the British Construction Industry*, Macmillan.

McGhie, W. (1981), 'The implications of project management' published in the proceedings of the Bartlett Summer School.

Seeley, I. H. (1975), *Building Economics*, Macmillan.

 CHAPTER 2

Andrews, J. (1982), unpublished college lectures (Construction Management), University College London.

Barnes, M. (1988), 'Construction project management', *Project Management*, **6**, (2), May, pp. 69–79.

Bunton, L. (1984), 'Avoiding disputes' *Building*, March, p. 25.

British Property Federation (1983), *Manual of the BPF System*, BPF.

CIOB (1982), *Project Management*, Chartered Institute of Building.

CIOB (1983), *Management Contracting*, Chartered Institute of Building.

CIRIA (1983), *Management Contracting*, CIRIA.

CIRIA (1984), *A Client's Guide to Management Contracts in Building*, CIRIA.

Davis, C. (1984), 'The BPF system – no going back', *Building*, June, pp. 28–9.

Davis, C. (1984), 'The BPF system – the system in action', *Building*, June, pp. 31–3.

Franks, J. (1984), *Building Procurement Systems*, CIOB.

Hillebrandt, P. M. (1977), *Economic Theory and the Construction Industry*, Macmillan.

Knoepel, H. and R. Burger (1987), 'Project organisation and contract management', *Project Management*, Butterworth & Co. Ltd, **5**,(4), November, pp. 204–8.

Morledge, R. (1987), 'The effective choice of building procurement method', *Chartered Quantity Surveyor*, July, p. 26.

Pennington, I. (1985), *A Guide to the BPF System and Contract*, CIOB.

RICS/IQS (1982), 'Does the client get what he wants?' *Quantity Surveyor*, pp. 86–8.

Royal Institute of British Architects (1988), 'Construction administration', *Architects' Job Book*, 1988.

Snowdon, M. (1980), 'Project management', *Quantity Surveyor*, November, pp. 2–4.

CHAPTER 3

AQUA Group (1990), *Tenders and Contracts for Building*, AQUA Group.

Banwell, H. (1964), *Placing and Management of Contracts for Building and Civil Engineering Works*, HMSO.

Brooks, D. (1986), 'Does public accountability achieve value for money', *Building Technology and Management*, March, pp. 11–14.

CONTRACT JOURNAL (1988), 'The right stuff; how does the client pick his builder?', *CJ*, October.

Doe (1982), *UK Construction Industry: A guide to methods of obtaining an industrial building in the UK*, Department of Environment.

Groaks, J. and Householder, J. (1990), 'Contractors' uncertainty and client intervention', *Habitat Int.*, **14**, (2–3), pp. 119–25.

Jones, W. G. (1979), *Contractor Selection: A guide to good practice*, CIOB Estimating Information Service.

Turner, D. F. (1979), *Quantity Surveying Practice and Administration*, Godwin.

CHAPTER 4

Andrews, J. (1989), unpublished college lectures (Construction Management), University College London.

Appelby, R. C. (1977), *Modern Business Administration*, Pitman.

Ball, M. (1988), *Rebuilding Construction*, Routledge.

Barnard, R. H. (1981), 'A strategic appraisal system for small firms', *Building Technology and Management*, September, pp. 21–4.

Blyth, D. and E. R. Skoyles (1984), 'A critique of resource control', *Building Technology and Management*, February, pp. 29–30.

Burch, T. (1982), *Absenteeism in the Building Industry*, CIOB Technical Information Service.

Burch, T. (1984), *Labour Turnover in the Construction Industry*, CIOB Technical Information Service.

Drucker, P. (1979), *The practice of management*, Pan.

Good, K. R. (1986), *Handling Materials on Site*, CIOB Technical Information Service.

Hambridge, B. W. (1982), *Productivity, Time and Cost*, CIOB Management Paper.

Handy, C. B. (1976), *Understanding Organisation*. Penguin.

Hillebrandt, P. M. (1984), *Analysis for the British Construction Industry*, Macmillan.

Horner, R. M. W. (1982), *Productivity, the Key to Control*, CIOB Technical Information Service..

Hillingworth, J. (1987), *The control of materials and waste*, CIOB Technical Information Service.

Hillingworth, J. (1988), *Handling of Materials on Site*, CIOB Technical Information Service.

Hillingworth, J. (1988), *Materials Management – Is it worth it?* CIOB Technical Information Service.

Ive, G. (1989), unpublished college lectures (Construction Economics), University College London.

Skoyles, R. (1984), *An Approach to Reducing Materials Waste on Site*, CIOB Technical Information Service.

Watts, B. K. R. (1982), 'Business and financial management' M & E monograph.

Whyatt, D. P. (1978), *The Control of Materials on Housing Sites*, CIOB Site Management Information Service.

 CHAPTER 5

Blain, B. C. R. (1978), *Work Study as an Aid to Estimating*, CIOB Estimating Information Service.

Braid, S. R. (1984), *Importance of Estimating Feedback*, CIOB Technical Information Service.

Burch, T. (1978/79), *Planning and Organisation Problems Associated with Confined Sites*, CIOB Site Management Information Service.

CIOB (1983), *Code of Estimating Practice*, Chartered Institute of Building.

Eastham, R. A. (1991), 'Tender thoughts: A requiem for condemned contractors', *Architect and Surveyor*, December, pp. 22–30.

Farrow, J. J. and D. K. Rutter (1990), *Performance Setting and Monitoring on Building Projects for Contractors*, CIOB Technical Information Service.

Fellows, R. F. and D. A. Langford (1980), 'Decision theory and tendering', *Building Technology and Management*, October, pp. 36–9.

Fletcher, A. L. (1981), *Attendance on Subcontractors*, CIOB Surveying Information Service.

Good, K. R. (1986), *Handling Material on Site*, CIOB Technical Information Service.

Griffiths, J. H. (1980), *Waste – An Estimator's Viewpoint*, CIOB Estimating Information Service.

Harrison, R. S. (1981), *Estimating and Tendering: Some aspects of theory and practice*, CIOB Estimating Information Service.

Massey, W. B. (1992), *Subcontractors During the Tender Period: An estimator's view*, CIOB Construction Papers.

NJCC (1983), *Code of procedure for two stage selective tendering*, NJCC.

NJCC (1991), *Code of procedure for single stage selective tendering*, NJCC.

Penn, H. (1978/79), *Estimating the plant element*, CIOB Estimating Information Service.

RICS (1975), *Definition of prime cost of dayworks carried out under a building contract*, Royal Institution of Chartered Surveyors.

Ryan, T. M. (1980), 'Plant cost estimating', *The Quantity Surveyor*, August, pp. 152–4.

Skoyles, E. R. (1982), *Waste and the estimator*, CIOB Technical Information Service.

Wakefield, N. (1984), 'Towards a true estimate', *Building*, June.

 CHAPTER 6

Ashworth, A. and D. A. Elliot (1986), *Price Books and Schedule of Rates*, CIOB Technical Information Service.

Ashworth, A. (1987), *The Computer and the Estimator*, CIOB Technical Information Service.

Atton, W. (1975), *Estimating Applied to Building*, Godwin.

Bailey, R. G. (1971), *Principles of Builders' Estimating and Final Account*, Crosby Lockwood.

Bentley J. I. W. (1987), *Construction Tendering and Estimating*, Spon.

Brook, M. (1988), *The Use of Spread Sheets*, CIOB Technical Information Service.

BRE DIGEST (1982), 'Material control to avoid waste', BRE.

BRE DIGEST (1981), 'Waste of building material', BRE.

Buchan R. D. (1991), *Estimating for Builders and Quantity Surveyors*, Newnes.

Chadwick, L. (1982), 'Materials management, profitability and construction industry', *Building Technology and Management*, June, p. 8.

CIOB (1981), *Estimating for Rehabilitation*, CIOB Estimating Information Service.
Everet, A. and A. King (1975), *Components and Finishes*, Mitchell.
Everet, A. (1975), *Materials*, Mitchell.
Griffiths, J. H. (1980), *Waste – An Estimator's Viewpoint*, CIOB Estimating Information Service.
Hall, D. S. M. (1972), *Elements of Estimating*, Batsford.
Harrison, B. A. (1984), *Pricing Drainage and External Works*, CIOB Technical Information Service.
Harrison, R. S. (1982), *Practicalities of Computer-assisted Estimating*, CIOB Technical Information Service.
Illingworth, J. (1987), *The Control of Materials and Waste*, CIOB Technical Information Service.
Illingworth, J. (1988), *Materials Management: Is it worth it?* CIOB Technical Information Service.
Miller, J. C. (1979), *Computer-aided estimating*, CIOB Technical Information Service.
Potter, D. (1982), *Computer-aided estimating*, CIOB Technical Information Service.
Sharp, J. A. A. (1981), *The Cost Estimate – A need for reconciliation*, CIOB Estimating Information Service.
Skoyles, E. R. (1982), *Waste and the Estimator*, CIOB Technical Information Service.
Skoyles, J. R. (1984), *An Approach to Reducing Materials Waste on Site*, CIOB Technical Information Service.
Smith, R. C. (1991), *Estimating and Tendering for Building Work*, Longman.
SMM7 (1988), *Standard Method of Measurement of Building Works*, Co-ordinated Project Information (CPI),.
Wainwright, W. H. and A. A. B. Wood (1991), *Practical Builders Estimating*, Stanley Thornes.

 CHAPTER 7

Bathurst, P. E. and D. A. Butler (1973), *Building Cost Control Techniques and Economics*, Heinemann.
Brook, M. (1991), *Safety Considerations on Tendering – Management's Responsibility*, CIOB Technical Information Service.
Cook, A. E. (1990), *The Cost of Preparing Tenders for Fixed Price Tenders*, CIOB Technical Information Service.

Cook, A. E. (1991), *Construction tendering*, Batsford.

Elliot, D. A. (1977), *Tender Patterns and Evaluation*, CIOB Estimating Information Service.

Fellows, R. (1991), 'Estimating – dinosaur or phoenix?', *Chartered Builder*, October, pp. 18–19.

Ferry, D. J. and P. S. Brandon (1980), *Cost Planning Buildings*, Granada.

Fine, B. 'Tendering strategy', in D. A. Turin (ed.), *Aspects of the Economics of Construction*, Godwin.

Flanagan, R. and G. Norman (1982), 'An examination of the tendering pattern of individual building contractors', *Building Technology and Management*, April, pp. 25–8.

Gray, C. (1983), 'Estimating preliminaries', *Building Technology and Management*, April, pp. 4–7.

Griffs, F. H. (1992), 'Bidding strategy: winning over key competitors, *Journal of Construction Engineering and Management*, **118**, (1), Paper no. 1234.

Hancock, M. R. (1990), *Theory of Markets and Price Formation in the UK Construction Industry*, CIOB Technical Information Service.

Henbsman, Z. and Ellis, R. (1992), 'Multiparameter, bidding system – innovation in contract administration', *Journal of Construction Engineering and Management*, **118**, (1), Paper no. 1436.

HM Customs & Excise (1990), *Value Added Tax: Construction industry*, VAT Leaflet 708/2/90.

Mudd, D. R. (1979), *Administration of Tender*, CIOB Estimating Information Service.

Murray, D. (1980), 'A degree of competition', *Quantity Surveyor*, October, pp. 187–8.

Nunn, D. (1991), 'When negative bids turn positively dangerous', *Contract Journal*, November, pp. 14–15.

RICS, *Building Cost Information Service (BCIS)*, Royal Institution of Chartered Surveyors, (Index no. 12), June 1993.

Seeley, I. H. (1978), *Building Economics*, Macmillan.

Senior, G. (1990), *Risk and Uncertainty in Lump Sum Contracts*, CIOB Technical Information Service.

Shash, A. A. (1993), 'Factors considered in tendering decisions by top UK contractors', *Construction Management and Economics*, **11**, pp. 111–18.

Skitmore, M. (1989), *Contract Bidding in Construction*, Longman.

Tassie, C. (1982), 'At the right price', *Building*, March, pp. 30–1.

CHAPTER 8

Coles, D. and Smith, G. (1982), 'Analysing the results of tenders', *Builders Trade Journal*, March, p. 33.

Cooke, A. E. (1990), *The cost of preparing tenders for fixed price contracts*, CIOB Technical Information Service.

Daniels, R. G. (1990), 'The cost of abortive tendering', *Quantity Surveyor*, pp. 129–30.

Davis, Langdon & Everest (1983), *Spon's Architects' and Builders' Price Book*, E. & F. N. Spon.

Fish, R. (1985), 'Tendering in a competitive market', *Chartered Quantity Surveyor*, August, **8**, p. 23.

Flanagan, R (1981), 'Change the system', *Building*, March, pp. 28–9.

Franks, J. (1990), 'Procurement in the 1990s', *Chartered Quantity Surveyor*, February, pp. 11–12.

Fryer, B. (1988), 'A new year look at the future of construction management', *Building Technology and Management*, February–March, pp. 33–5.

Hancock, M. (1978), 'The selection of contractual arrangements', *Chartered Quantity Surveyor*, July, pp. 12–13.

Handy, J. J. (1992), *Germany – A Challenge for the Estimator*, CIOB Construction Papers.

Lange, J. E. and D. Q. Mills (1979), *The Construction Industry*, Lexington Books.

Millwood, M. (1983), 'Changing methods of placing contracts', *Chartered Quantity Surveyor*, p. 177.

Morris, W. G. (1988), 'Management thinking and the building industry', *Building Technology and Management*, June–July, pp. 30–4.

Stockdale, D. J. (1982), *Estimating in the Middle East*, CIOB Technical Information Service.

Stockdale, D. J. (1982), *Preparation of Cost Estimates Using the North American Index*, CIOB Technical Information Service.

Wainwright, W. H. and A. A. B. Wood (1985), *Variation and Final Account Procedure*, Hutchinson.

INDEX

The Noah Project

The Secrets of Practical Project Management

Ralph L Kliem and Irwin S Ludin

This book is a novelization of project management. The characters and events are fictitious; however the techniques, tools, and circumstances described in each chapter are real for just about every project in any environment, from technical to financial. The scenes explain project management from the vantage point of David Michaels, a young executive working for a private zoo. He must manage the dismantling of the zoo through to a successful conclusion. He has little idea how to go about such a task until he meets Noah...

David encounters common pitfalls such as failure to achieve targets on time, budgeting restrictions, an already unreasonable schedule cut back even further, and of course the inevitable staff conflict. In his moments of crisis Noah forces David to think for himself, thereby encouraging the reader to do the same.

The authors have chosen the setting intentionally to show how anyone in any organization can put the methods and concepts of project management to use. The book also includes a 'model' project manual which can be adapted easily to the reader's own projects. Anyone looking for an enjoyable introduction to the secrets of project management will find it in *The Noah Project*.

1993   208 pages   0 566 07439 7

Gower

Project Management
Fifth Edition

Dennis Lock

Project management is the function of evaluation, planning and controlling a project so that it is finished on time, to specification and within budget. It uses a family of techniques which can be practised with profit whether the project is worth £1000 or £1m.

Dennis Lock's book explains and demonstrates these techniques in action. It covers project management from initial appraisal to close-down, using methods ranging from simple charts to powerful computer systems, and with every subject explained using step-by-step illustrations and case studies. When it first appeared in 1968, it was acclaimed as a pioneering work: it is still the standard work on its subject, for managers and students alike.

This fifth edition has been completely revised and updated since the publication of the previous edition in 1988. Several new illustrations have been introduced and many of the existing ones replaced. Whilst the logical sequence of topics throughout the book remains the same the new edition now contains three additional chapters: on project management organization, on commercial management and on advanced procedures and systems. The result is an improved and strengthened edition that is completely up to date with the latest practice and technology. It is quite simply the standard text for anyone in industry who is interested in, or responsible for, projects.

Project Management is an Open University set text.

1992   542 pages   0 566 07339 0 Hardback   0 566 07340 4 Paperback

Gower